GLOBAL SOIL SECURITY

PROCEEDINGS OF THE GLOBAL SOIL SECURITY 2016 CONFERENCE, PARIS, FRANCE, 5–6 DECEMBER 2016

Global Soil Security

Towards More Science-Society Interfaces

Editors

Anne C. Richer-de-Forges
INRA, France

Florence Carré
INERIS, France

Alex B. McBratney
University of Sydney, Australia

Johan Bouma
Wageningen University, The Netherlands

Dominique Arrouays
INRA, France

CRC Press is an imprint of the
Taylor & Francis Group, an **informa** business

A BALKEMA BOOK

Cover artwork: Anne C. Richer-de-Forges, INRA, France

CRC Press/Balkema is an imprint of the Taylor & Francis Group, an informa business

© 2019 Taylor & Francis Group, London, UK

Typeset by V Publishing Solutions Pvt Ltd., Chennai, India

All rights reserved. No part of this publication or the information contained herein may be reproduced, stored in a retrieval system, or transmitted in any form or by any means, electronic, mechanical, by photocopying, recording or otherwise, without written prior permission from the publisher.

Although all care is taken to ensure integrity and the quality of this publication and the information herein, no responsibility is assumed by the publishers nor the author for any damage to the property or persons as a result of operation or use of this publication and/or the information contained herein.

Published by: CRC Press/Balkema
　　　　　　　Schipholweg 107C, 2316 XC Leiden, The Netherlands
　　　　　　　e-mail: Pub.NL@taylorandfrancis.com
　　　　　　　www.crcpress.com – www.taylorandfrancis.com

ISBN: 978-1-138-09305-8 (Hbk)
ISBN: 978-1-315-10707-3 (eBook)

Table of contents

Foreword	vii
Committees	ix
Preface	xi
Acknowledgements	xiii

Introduction to global soil security

How the soil security concept can pave the way to realizing some soil related UN-SDG's *J. Bouma*	3
The concept of soil security *A.B. McBratney, M. Moyce, D. Field & A. Bryce*	11
Approach to valuing soil ecosystem services with linking indicators *D.K. Bagnall, C.L.S. Morgan, R.T. Woodward & Wm.A. McIntosh*	19

Soil security and policy

Global soil security for future generations *L. Montanarella & P. Panagos*	29
Contribution of knowledge advances in soil science to meet the needs of French state and society *V. Antoni, H. Soubelet, G. Rayé, T. Eglin, A. Bispo, I. Feix, M.-F. Slak, J. Thorette, J.-L. Fort & J. Sauter*	33

Soil security and mapping

GlobalSoilMap and the dimensions of Global Soil Security *D. Arrouays, A.C. Richer-de-Forges, A.B. McBratney, B. Minasny, M. Grundy, N. McKenzie, Z. Libohova, P. Roudier & J. Hempel[†]*	43
DSM for soil functions: A preliminary example using spatialized BBNs in Scotland *L. Poggio & A. Gimona*	47
Soil systems—a soil survey approach to soil security *Z. Libohova, P. Schoeneberger, D. Wysocki, S. Wills, C. Seybold & D. Lindbo*	53
Soil information system: The pathway to soil and food security in Haiti *C. Kome, P. Reich, J. Lene, Z. Libohova, S. Monteith, P. Finnell, S. McVey, L. Scheffe, S. Southard, S. Bailey, T. Rolfes, N. Jones & M. Matos*	57

Soil security and practitioners

Overview of tillage practices and correlations with other practices in France: An analysis of the agreste survey (2011) *N. Cavan, J. Labreuche, A. Wissocq, F. Angevin & I. Cousin*	65
Balancing decisions for urban brownfield regeneration: People, planet, profit and processes *L. Maring, F.L. Hooimeijer & J. Norrman*	73

Extension of irrigation in semi-arid regions: What challenges for soil security? Perspectives from a regional-scale project in Navarre (Spain)   79
R. Antón, I. Virto, J. González, I. Hernández, A. Enrique, P. Bescansa, N. Arias, L. Orcaray & R. Campillo

Connectivity and raising soil awareness

Soil security to connectivity: The what, so what and now what   91
D. Field

The non-anthropocentric value of soil and its role in soil security and Agenda 2030   99
P.E. Back, A. Enell & Y. Ohlsson

Soil awareness in Italian high schools: A survey to understand soil knowledge and perception among students   107
M.C. Moscatelli, S. Marinari & S. Franco

Opportunities for enhancing soil health   113
C.W. Honeycutt, S.R. Shafer, S.R. Jones & B.F. Rath

The soil certificate—a Flemish tool helping raise awareness about soil   119
J. Ceenaeme, G. Van Gestel, N. Bal & W. Van Den Driessche

Soil security and research needs

Which R&D needs for a sustainable soil management and land use?   127
M.C. Dictor, V. Guérin & S. Bartke

Conclusion

The 2nd global soil security conference—conclusions and prospects   133
A.C. Richer-de-Forges, D. Arrouays, F. Carré, J. Bouma & A.B. McBratney

Author index   137

Foreword

THE NEED FOR GLOBAL SOIL SECURITY

The term "soil security" refers to protection, management and restoration of the world's soil resources to strengthen ecosystem services for human wellbeing and nature conservancy. In this context, the term soil security is homologous to similar terms used for protection and management of other critical natural resources such as water security, climate security, biodiversity security, and energy security. The strategy is to protect, manage and restore soil to meet the demands of the world's growing (7.6 billion in 2018 and projected to reach 9.8 billion in 2050 and 11.2 billion in 2100) and increasingly affluent population especially in densely populated and resource-scarce countries. Advancing soil security is critically important because on it depends the security of water, climate, biodiversity, and energy. An effective approach to addressing global issues of the 21st century, not only biophysical but also socioeconomic and cultural/aesthetical along with the dire need of achieving nation/regional and global peace, depends on managing and improving soil security at all levels. This interconnectivity is also in accord with the UNCCD's goal of achieving "Land Degradation Neutrality", although we would hope to do better than a zero-sum game. Furthermore, such an inter-dependence (nexus) is also essential to advancing the Sustainable Development Goals (SDGs) of the UN or the so-called "Agenda 2030." For example, achieving soil security is essential to advancing SDG #2 (zero hunger), #3 (good health and wellbeing), #6 (clean water), #13 (climate action), and #15 (life on land). Soil perhaps will be recognized with its own SDG eventually, e.g., soil action. Indeed, peace, prosperity, political stability and harmonious living in symbiosis with nature all depend on soil security. The importance of the concept of soil security is also highlighted by the fact that "health of soil, plants, animals, people, and ecosystems is one and indivisible." Thus, healthy ecosystems, dependent upon healthy and secure soil resources, necessitate that the scientific knowledge of soil management and enhancement be translated into action through appropriate policy interventions so that global concepts such as SDGs and Land Degradation Neutrality can be achieved at local, regional, national, and global scales, with global soil security being an essential factor in such accomplishment.

The book is based on the papers presented at the 2nd Soil Security Conference held in Paris from 5–6 December 2016. The first book in the series; "Damien J. Field, Cristine L.S. Morgan and Alex B. McBratney (Eds). 2017. Global Soil Security, Springer, 463 pages," comprised 43 chapters. The present volume is also edited by highly knowledgeable professionals (Anne Richer-de-Forges, Florence Carré, Alex B. McBratney, Johan Bouma, and Dominique Arrouays) and comprises chapters contributed by eminent soil scientists practitioners and policymakers from around the world. The second book, published by CRC Press/Balkema, Taylor & Francis Group, complements and updates the material presented in the first book. The information presented herein is of interest to researchers, students, farmers, land managers, development organizations, and policymakers. It addresses the basic concepts, management-induced differences in soil processes and properties, knowledge gaps and researchable priorities, policy issues and action plan.

Rattan Lal
President, International Union of Soil Sciences,
Distinguished University Professor of Soil Science,
The Ohio State University,
Columbus, USA
20th July 2018

Committees

ORGANIZING COMITTEE

Dominique Arrouays (INRA)
Florence Carré (INERIS)
Céline Collin Bellier (AFES)
Adila Omari (AFES)
Anne Richer-de-Forges (INRA)

SCIENTIFIC COMMITTEE

Alex B. McBratney (*University of Sydney, Australia*)
Blair McKenzie (*The James Hutton Institute, UK*)
Damien Field (*University of Sydney, Australia*)
Isabelle Feix (*ADEME, France*)
Jean-Francois Soussana (*INRA, France*)
Allan Lilly (*The James Hutton Institute, UK*)
Christian Valentin (*IRD, France*)
Claire Chenu (*AgroParisTech, France*)
Cristine Morgan (*University of Texas, USA*)
David Bendz (*SGI, Sweden*)
Johan Ceenaem (*OVAM, Belgium*)
Johan Bouma (*Wageningen University & Research, The Netherlands*)
Luca Montanarella (*JRC, Italy*)
Ronald Vargas (*FAO, Italy*)

Preface

Some 140 scientists, policy advisers, investors, and citizens from 25 countries met in Paris, France, on the 5th and 6th December 2016 to discuss the need for soil security. The Symposium was organised jointly by the French National Institute for Agricultural Research (INRA), the French National Institute for Industrial Development and Risk (INERIS), and the French Soil Science Society (AFES). Numerous institutes and organisations, the logos of which appear in the acknowledgments section, generously supported the symposium.

The symposium was opened by a talk from the French Minister of Agriculture, Mr Stéphane Le Foll.

Attendees participated in an open discussion focused on each participant's perspective on how to achieve soil security, that involved scientific, economic, industrial and political engagement informing users, citizens and policy makers, with the objective of implementing appropriate actions. The contributions to this book address the five dimensions of soil security: capability, condition, capital, connectivity and codification.

We hope that *Global Soil Security: Towards more science-society interfaces* will stimulate the development of the soil security concept and its effective implementation in all parts of the world. Action is needed urgently to address global issues such as food security, water security, climate change and biodiversity protection.

The discussion of Global Soil Security will continue with a focus on soil security and planetary health at the 2018 Global Soil Security Conference in Sydney, December 4–6, 2018.

Anne Richer-de-Forges
Florence Carré
Alex B. McBratney
Johan Bouma
Dominique Arrouays
Paris, July 2018

Global Soil Security – Richer-de-Forges et al. (Eds)
© 2019 Taylor & Francis Group, London, ISBN 978-1-138-09305-8

Acknowledgements

The conference was organized in Paris in 2016 by AFES, INRA and INERIS.

ORGANIZED BY

SPONSORED BY

Introduction to global soil security

How the soil security concept can pave the way to realizing some soil related UN-SDG's

J. Bouma
Wageningen University, Wageningen, The Netherlands

ABSTRACT: The seventeen Sustainable Development Goals, approved by the General Assembly of the United Nations in 2015, present an attractive focus for soil research at a time when societal relevance of research is increasingly questioned by the policy arena and by societal pressure groups. Soil science cannot address the SDG's in isolation but the profession is in an excellent position as pathways towards SDGs related to food, water, climate and biodiversity can only be defined with substantial input by soil science. Interdisciplinarity, thus required, calls for defining soil contributions to Ecosystem Services, that, in turn, are essential elements to contribute to SDGs. Soil functions define such contributions, using functional soil characteristics, but they can become more effective when also articulated in terms of the 5C's, as introduced in the context of the Soil Security concept. A storyline is introduced, linking the 5C's in a logical sequence to each one of the soil functions thereby improving the effectivity of soil input into the SDG discours.

1 INTRODUCTION

The concept of sustainable development was first introduced in 1988 in the Brundtland report emphasizing the need to not only consider economic factors when trying to achieve sustainable development but also social and environmental aspects. Though new and attractive, the concept was somewhat abstract and difficult to translate into operational criteria. The introduction of 17 specific Sustainable Development Goals (SDGs) by the United Nations in 2015 has changed this and is providing well defined goals presenting a challenge to both the policy and research arena. At least five of the seventeen SDG's have a direct relation with soils (Table 1).

Three aspects are important when considering soil research to be focused on SDGs:

Table 1. Five SDGs with particular relevance for soils.

2. End hunger, achieve food security and improve nutrition and promote sustainable agriculture (FOOD)
3. Ensure healthy lives and promote well being for all at all ages. (HEALTH)
6. Ensure availability and sustainable management of water and sanitation for all. (WATER)
13. Take urgent action to combat climate change and its impacts. (CLIMATE)
15. Protect, restore and promote sustainable use of terrestrial ecosystems, sustainably manage forests, combat desertification and halt and reverse land degradation and halt biodiversity loss. (ECOSYSTEMS)

i. soil science cannot be effective on its own and needs to engage with other disciplines, such as agronomy, hydrology, climatology and ecology, to adequately define ways to reach certain SDGs. This can be achieved by focusing first on Ecosystem Services (ES) that, in turn, provide a contribution towards SDG's. (Table 2);
ii. SDGs are defined in an action mode: "end", "ensure", "take action", "protect" and "restore". Talking time is over: action-oriented research is needed; (iii) data, information and knowledge

Table 2. Ecosystem Services (ES) with an important soil component according to Dominati et al. (2014).

Provisioning services
  1. Provision of food, wood and fibre.
  2. Provision of raw materials.
  3. Provision of support for human infrastructures and animals.
Regulating services
  4. Flood mitigation
  5. Filtering of nutrients and contaminants
  6. Carbon storage and greenhouse gases regulation
  7. Detoxification and the recycling of wastes
  8. Regulation of pests and disease populations
Cultural services
  9. Recreation
  10. Aesthetics
  11. Heritage values
  12. Cultural identity

are relevant in as much as they are relevant in the context of being effective in achieving SDGs, and (iv) relations among SDGs are important, avoiding separate approaches for each one of them. Production of healthy food can go together with a good quality of ground- and surfacewater using precision techniques when applying agrochemicals, while increasing the organic matter content of soils by management which is favorable for climate mitigation and ecological quality. A systems approach for soil management is therefore needed when studying SDGs.

2 THE NEED FOR INTERDISCIPLINARITY

The soil science discipline has a rather self-centered history, not unlike many other scientific disciplines. Soil surveys have reported soil suitabilities or limitations for a range of land uses. The introduction of land evaluation by FAO (1976, 2007) broadened the soil concept to land, thereby also considering climate, geology and hydrology (e.g. Bouma et al, 2012). Still, land suitabilities were defined based on land qualities and associated land characteristics, in fact ignoring essential agronomic, economic and social aspects. But soil survey interpretations and land evaluations have been and still are quite useful and effective when considering large, relatively unexplored areas of land. However, modern applications require soil scientists to wonder what kind of soil information is important when participating in interdisciplinary programs with agronomists, climatologists and hydrologists, not to mention economists and sociologists. Bouma (2014) and Keesstra et al, (2016) have therefore emphasized the need to consider soil contributions to ecosystem services, that, in turn, contribute to studies focusing on realizing SDGs (Figure 1).

The paradigm shift from providing suitability judgements rather than defining the most effective soil input into interdisciplinary programs has more serious consequences than is generally realized. One reflex is to ask what other disciplines need and this translates into a function as service provider which is deadly from a scientific point of view. The way databases are now established in the soil science discipline, applying modern digital information technology, tends to go in that direction when soil scientists are not directly involved in modeling exercises made by others. A more pro-active approach focuses not on just providing soil data that remains rather abstract to outsiders but on information on soil functions indicating specific contributions that soil science can deliver. Functions were defined by EC (2006) as shown in Table 3 and indicated in Figure 1.

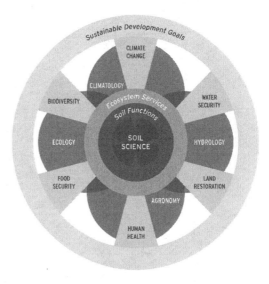

Figure 1. Diagram illustrating relations between soil functions, ecosystem services and SDGs (after Keesstra et al, 2016).

Table 3. The seven soil functions (SF) as defined by the European Commission (EC, 2006).

1 Biomass production, including agriculture and forestry.
2 Storing, filtering and transforming nutrients, substances and water.
3 Biodiversity pool, such as habitats, species and genes.
4 Physical and cultural environment for humans and human activities.
5 Source of raw material.
6 Acting as carbon pool.
7 Archive of geological and archaeological heritage.

This, however, leaves the question as to how these "functions" can be translated info specific information that is important for interdisciplinary approaches focusing on ESs and the SDGs. Here, the Soil Security concept, defining "The Five C's", can play a central role in making soil input more interesting and relevant. (McBratney et al, 2004, Field et al, 2016) For each soil function condition, capability, capital, connectivity and codification can be defined, and can be combined into a "storyline" that translates the, as such, still rather broadly and abstractly defined soil functions in a real-world discours.

The storyline could go like this:

Considering a given type of soil and each of the seven soil functions, how and by whom is the soil being used and managed? What are their questions and goals? Who passes judgements? ("connectivity"). What is its "condition" in terms of its contribution towards ecosystem services and what contributions might be potentially possible? ("capability"). How does this soil compare with other soils in terms of its contributions ("capital") and are its condition and capability properly addressed in societal and policy legal frameworks ("codification").

This storyline will form the basis for discussions on inter- and transdisciplinarity in the following sections.

3 ORIENTATION ON RESULTS REQUIRES CONNECTIVITY

As may be expected, the SDG's are formulated in a goal-oriented manner. General statements as to the need for improved communication between science and society, sustainable development, inter- and transdisciplinarity, more funds for research etc. that are all too often found in general review or policy papers, will not satisfy anymore. The changing relations between science and society imply the urgent need for a drastic re-evaluation of scientific practices.

The often used characterization of current political and social conditions as being "post-truth" and "fact-free" is misleading. Different ideological visions on society have their own "truth" which is supported by their own "alternative facts", requiring selective framing of facts that are reported in the press or in databases. "Facts" that don't fit within a particular "truth" are denied or, worse, falsified. When scientists talk about "truth" they refer to the scientific method where "facts" are generated by well defined procedures defining a problem, a hypothesis, application of existing reproducible methods or development of new methods and testing of results by statistical methods. Scientists know that an absolute "truth" does not exist and, ideally, they falsify their hypotheses to move forward, increasing their understanding of a very complex world step by step. Of course, societal development has resulted from scientific advances over the years but many 21th citizens, glued to their websites and influenced by populist notions, prefer their own "truth" and "facts". In fact "post-truth" is: "post-our truth" and "fact-free" is: "free from our facts", where "our" refers to the scientific community. Science has to re-connect and this also applies to soil science! The time that grateful stakeholders and policy makers looked up in admiration to scientists in the hope to replenish their desperate knowledge-starved existence, is over. "Connectivity" is more important than ever.

The literature shows many examples of successful interactive research, that can be illustrated by the diagram of Figure 2 (Bouma et al, 2008).

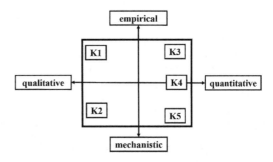

Figure 2. Knowledge diagram, as discussed in text (after Bouma et al, 2008).

Two axes range from empirical to mechanistic vertically and from qualitative to quantitative horizontally. K1 represents stakeholder knowledge; K2 represents a better understanding of underlying mechanisms and qualifies as expert knowledge. K3 to K5 are in the scientific arena where K3 is still rather empirical (e.g. application of regressions analysis) and K4 and K5 represent increasingly quantitative research, using sophisticated techniques and models. K5 research qualifies best for publications in international refereed scientific journals. As scientific careers depend on publication records, this presents a problem because stakeholders and policy makers may see K4-K5 research in terms of answering questions that have never been asked.

To more and better engage stakeholders, a start of research at K1 level is preferable. Not only asking what the questions are but, particularly, which goals are being pursued. The SDGs form a suitable set of goals. The latter provides more operating room than strictly focusing on questions. Jointly moving on to K2 can represent a joint-learning procedure that can be intensified by moving up to K3, K4 and even K5. Bouma et al, (2015) analysed seven already published case studies in the Netherlands and Italy and found that in three of them existing K3 knowledge was adequate to answer problems raised. But in the remaining cases K5 research was needed. Taking the K1 to K5 route in dialogue with stakeholders the research process becomes more transparent and the need for new research can be demonstrated more clearly than just by asking for more funds.

The K1-K5 approach is suggested to cement "connectivity". It will take more time than just starting with K4-5 research but it will pay off in more engagement and commitment of stakeholders and policy makers. It may be advisable to approach the policy makers through the stakeholders, or, rather voters. Note that the arrows in Figure 2 point both ways: when following the K1 to K5 approach, knowledge is also communicated effectively down the line.

As mentioned, emphasis on goals rather than questions allows articulation of the particular "truth" of certain stakeholders and research can show what the economic, social and environmental consequences are of any particular "truth", thereby defining the degree of sustainability being obtained in terms of SDGs. Stakeholders will decide what to do, not the scientist. Alternative "truths", reflecting opinions of other stakeholder groups can also be defined in a scenario analysis, allowing comparisons between alternative scenario's trying to satisfy SDGs.

Working jointly in the K1-K5 mode, often implies that alternative "truths" are developed during the process as a result of joint learning and that represents a more effective process of transfer of scientific research than a more traditional top-down approach.

4 DATA, INFORMATION AND KNOWLEDGE AS A MEANS TOWARDS A PURPOSE

The soil science profession has been quite effective in their response to requests by other disciplines to provide soil data for their global and regional simulation models. This was particularly true for climate—change studies by the International Panel for Climate Change. Sanchez et al, (2009) pointed out that deriving data from existing soil surveys had some limitations and that systematically providing soil characteristics for grid data by digital soil mapping (McBratney et al, 2011) had much promise. Such data, as used in modeling studies on climate, water management and agronomic practices would be welcome as they could be translated into model parameters, applying pedotransfer functions (e.g. Bouma, 1989, Vereecken et al, 1992). Several world-wide programs are now providing this type of data (Arrouays et al, 2014, Hengl et al, 2014). These databases also indicate soil classifications that are considered to be representative for any particular grid point (IUSS Working Group WRB, 2015, Soil Survey Staff, 2010). Other global databases are more directly linked to soil surveys (FAO et al, 201, Batjes, 2016, Stoorvogel et al, 2017a, b). A minimum dataset has been determined in order to determine the most important soil functional properties and consists of (McBratney et al, 2017):

Depth to rock, plant exploitable depth, organic carbon, pH, Clay, Silt, Sand, coarse fragments, ECEC, Bulk density, Bulk Density of the fine earth, available water capacity (AWC), Electrical Conductivity (EC).

The step from these properties to SDG's is, of course, still large. Some properties, such as pH, depth to rock and EC directly describe functional soil properties that are important to define Ecosystem Services. Others, such as texture, organic carbon and bulk density derive their value in feeding pedotransfer functions, defining hydrological soil properties such as hydraulic conductivity and moisture retention (e.g. Vereecken et al, 1992). The AWC is a static value defining the volume of water held between two pressure heads corresponding with "field capacity" (0.3 bar) and "wilting point" (15 bar). Simulation models of the soil water regime, following the "tipping-bucket" representation, use AWC but this represents a rather strong simplification because the water that is really available to plants is governed by the climate and by more realistic representations of the water extraction process by plants. (e.g. Bonfante and Bouma, 2015).

As stated above, databases also include the soil type that is dominant in each grid cell. This can be the soil series (Soil Survey Staff, 2010) or the (IUSS-WRB, 2015). The defining characteristics of each soil type have to be clear allowing unequivocal distinctions as was illustrated by Bonfante and Bouma (2015) for six soil series in the area in Italy. Distinctions of soil types allows their use as class-pedotransfer functions (Bouma, 1989): data obtained for a given soil type can be assumed to also apply when a similar soil type is found at a new location, assuming that the climate is identical. This is not possible when just 13 separate soil characteristics are present. Also the 5 C's of Soil Security can be better associated with a given soil type than with a given grid point. The procedure to link soil functions, expressed by the 5C's as defined in Soil Security, to ESs and SDGs will now be illustrated for SDG2.

5 LINKING SF WITH ES WHEN AIMING FOR AN SDG: SDG2 AS AN EXAMPLE

SDG2 (Table 1) covers not only food security but also sustainable agriculture, implying a consideration of economic, social and environmental aspects associated with agricultural production. This is quite a package. Confining attention here to biomass production (soil function 1) links are found with ES 1 (provision of food) but also with ES8 regulating pests and diseases (Table 2). But when production of food is considered, agronomists will also deal with crop selection, including varieties and crop rotations as a function of climate conditions. Thus, the functional soil properties, mentioned earlier, can be presented to agronomists and jointly an analysis can be made as to relevant ESs and, after that the SDG. The latter represents the ultimate goal of the SDG analysis that will also have to consider socio-economic conditions, looking at the entire foodchain, logistics, and markets. After all, articulating an ES is only one of many contributions that will in the end allow a satisfactory pathway to a given SDG, in this case SDG2.

This chapter proposes to extend soil functions (here: function 1) beyond a set of soil functional properties by including the five C's of the Soil Security concept (e.g. Field et al, 2017), following the storyline presented in section 2. Such a more detailed description of any given soil function is more informative when defining ESs and SDGs.

So, continuing the example of SDG2 (restricted to food safety), Soil function 1 will not be limited to mentioning biomass production, but it will also describe current conditions at a given site where a given type of soil occurs, indicate its capability and the capital it represents. Importantly, information as to who uses and manages the soil and which outsiders affect the way the soil is being used as well as which forms of codification and legislation affect soil use is of particular interest for the broad SDG analysis.

6 OTHER SOIL FUNCTIONS

Soil function 2 (Table 2) has a special character. The capacity of a soil to store, filter and transform nutrients is a clear function of several of the soil functional properties mentioned in section 4. One limitation of the % clay property is lack of information on clay minerals present, which is part of the soil type description. Kaolinites, for example, have a very low CEC, montmorillonites have very high values. Also,% C does not distinguish between different types of organic matter with different adsorption capacities. But in addition to this, management plays an important role in determining the filtering capacity of soil. Pathogenic viruses were completely removed when spiked wastewater was applied to sand columns at rates of 5 cm/day while viruses flowed through a column of 60 cm length at flow rates of 50 cm/day. The difference was due to lower flow rates through finer soil pores at the lower application rate (Figure 3). Different processes operated in well structured B-horizon in a silt loam soil where fecal coliforms and fecal streptococci were removed in a 60 cm long soil column at the very low flow rate of 5 mm/day while they moved rapidly through planer voids between the peds at a higher rate as a result of bypass flow (Figure 4).

Functions 3 and 6 have a relation as higher organic matter contents of soil correspond with a higher biodiversity. These functions are highly relevant if only because of the recent "4per1000" proposal accepted at the Paris climate conference in 2015. As every soil type is characterized by unique physical, chemical and biological processes, most effective management procedures to increase the organic matter content are not only different for

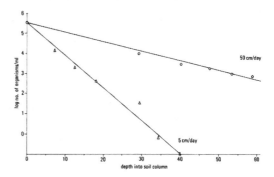

Figure 3. Removal of pathogenic viruses at two flow rates through a sand column (from Bouma, 1979).

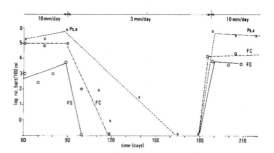

Figure 4. Removal of pathogenic bacteria at two flow rates through a well structured silt loam column (from Bouma, 1979).

different soils but are also more effective in certain type of soils as compared with others. Bouma and Wosten (2016) compared the range of organic matter contents in a class of clay soils in the Netherlands and related that to management, particularly in terms of "good conditions" and "greening" as defined in the EU Common Agricultural Policy. Simulation of organic matter dynamics in soils is difficult and a plea has therefore been made to sample fields where a given type of soil occurs but where management has been different over the years, resulting in different organic matter contents: phenoforms versus genoforms, as discussed elsewhere. (e.g. Bouma et al, 2017).

Soil functions 4 and 7 deal with cultural aspects where soils are considered in a broader regional context. But also for these functions, the 5C's are highly relevant.

7 AVOID A FOCUS ON SINGLE SDG'S

As mentioned above, sustainable forms of agriculture don't only focus on biomass production but also protect the quality of ground- and surfacewater, while making an attempt to increase the organic matter content of soils thereby contributing to climate change mitigation and to biodiversity preservation. This covers SDG's 2, 6, 13 and 15 in one stroke, while vital agriculture can significantly contribute to reducing poverty (SDG1), healthy lives (SDG3), gender equality in developing countries (SDG5), economic growth (SDG 8) and promotion of peaceful societies, particularly in developing countries (SDG 16). This may all seem rather redundant but "framing" soil science activities in the context of SDGs represents an important element of communication to the outside world. Bouma et al, (2015) analysed six already published case studies in the Netherlands and in Italy and could "frame" results of each of the studies in terms of several SDG's. The same applied to a study of alternative dairy systems in the Northern Frisian Woods in the Netherlands, where application of less fertilizers and improved land management by farmers rather than contractors, resulted in better water quality (SDG6), a higher net income due to lower costs (SDGs 1, 3 & 12) and significantly higher soil organic matter contents (SDG 13) (Dolman et al, 2014; De Vries et al, 2015). Integration of different SDGs is also enhanced by applying techniques of precision agriculture where plant nutrients, water and biocides are only applied when needed during the growing season (McBratney et al, 2005; Stoorvogel et al, 2015).

8 FROM POINTS TO AREAS OF LAND

The various recently developed databases, reviewed above, provide data for grid points of varying dimensions. So far in this chapter the discussion focused on SFs, ESs and SDGs as they can be defined for point data. But SDGs in the real world relate to areas of land that can vary from a single field to a watershed or even much larger areas. Classical soil survey interpretations provide interpretations for areas of land, shown as mapping units on a given soil map. The assumption is made that a so-called "representative soil profile" adequately represents the soils within that particular mapping unit, thereby implicitly ignoring spatial variability. Many studies have been made on spatial variability, some of them applying geostatistics, and a discussion of this work is beyond the scope of this text. However, sometimes rapid results are needed for planning purposes and then modern interpretation techniques, using simulations, are combined with the classical visual representations based on patterns of the soil map. Bonfante and Bouma (2015) reported a study on the irrigation potential in an area in Southern Italy, also considering the introduction of new cultivars of maize. Their conclusions were highly relevant for planning purposes because soils in the area had significantly different capabilities. Still, the assumption was made that mapping units were homogeneous and soils could be represented by a single "representative" soil profile. Future studies are needed to show whether or not modern spatial variability studies would change conclusions reached.

9 CONCLUSIONS

1. The SDG's provide an attractive focal point for soil studies if only because of its communication potential when presenting specific case studies.
2. Considering the 5C's of the Soil Security concept when defining soil functions makes the latter much more attractive as input into interdisciplinary studies.
3. Presenting the 5C's in terms of a logical storyline can be helpful to integrate the Soil Security concept into soil science practice.

REFERENCES

Arrouays, D., Grundy, M.G., Hartemink, A.E., Hempel, J.W., Heuvelink, G.B.M., Young Hong, S., Lagacherie, P., Lelyk, G., Mc Bratney, A.B., Mc Kenzie, N.J., Mendonca-Santos, M., Minashy, B., Montanarella, L., Odeh, I.O.A., Sanchez, P.A., Thompson, J.A., Zhang, G.L. 2014. Global Soil Map: Toward a fine-resolution grid of soil properties. Advances in Agronomy 125: 93–134.

Batjes NH 2016. Harmonised soil property values for broad-scale modelling (WISE30sec) with estimates of global soil carbon stocks. *Geoderma* 269, 61–68.

Bonfante, A. and J. Bouma. 2015. The role of soil series in quantitative Land Evaluation when expressing efects of climate change and crop breeding on future land use. Geoderma 259–260, 187–195.

Bouma, J., 1979. Subsurface applications of sewage effluent. In: M.T. Beatty, G.W. Petersen & L.D. Swindale (ed.), Planning the uses and management of land. Agronomy 21, ASA-CSSA-SSSA, Madison Wisc. USA. p. 665–703.

Bouma, J., 1989. Using soil survey data for quantitative land evaluation. Advances in Soil Science, Vol. 9. B.A. Stewart (Ed.): Springer Verlag, New York: 177–213.

Bouma, J. 2014. Soil science contributions towards Sustainable Development Goals and their implementation: linking soil functions with ecosystem services. J. Plant Nutrition and Soil Sci. 177 (2): 111–120.

Bouma, J., 2015. Engaging soil science in transdisciplinary research facing wicked problems in the information society. Soil Sci. Soc. Amer. J. 79: 454–458. (doi: 10.2136/sssaj2014.11.0470).

Bouma, J. and J.H.M Wösten. 2016. How to characterize: good and "greening" in the EU Common Agricultural Policy (CAP): the case of clay soils in the Netherlands.

Soil Use and Manag. 32(4), 546–552. (doi: 10.1111/sum 12289).

Bouma, J., J.A.de Vos, M.P.W. Sonneveld, G.B.M. Heuvelink and J.J. Stoorvogel. 2008. The role of Scientists in multiscale land use analysis: lessons learned from Dutch communities of practice. Advances in Agronomy 97: 177–239.

Bouma, J., J.J. Stoorvogel and W.M.P. Sonneveld. 2012. Land Evaluation for Landscape Units. Handbook of Soil Science, Second Edition. P.M. Huang, Y. Li and M. Summer (Eds). Chapter 34. P. 34-1 to 34–22. CRC Press. Boca Raton.London. New York.

Bouma, J., C.K. wakernaak, A. Bonfante, J. Stoorvogel and L.W. Dekker. 2015. Soil science input in Transdisciplinary projects in the Netherlands and Italy. Geoderma Regional 5, 96–105. (http://dx.doi.org/10.1016/j.geodrs.2015.04.002).

Bouma, J., M.K. van Ittersum, J.J. Stoorvogel, N.H. Batjes, P. Droogers and M.M. Pulleman. Soil Capability: exploring the potenials of soils. In: Global Soil Security 2017 (D.J. Field, C.L.S. Morgen and A.B. McBratney (eds), 27–44. Springer International, Switzerland.

De Vries, W., J. Kros, M.A. Dolman, Th.V. Vellinga, H.C. de Boer, M.P.W. Sonneveld and J. Bouma, 2015. Environmental impacts of innovative dairy farming systems aiming at improved internal nutrient cycling: a multi-scale assessment. Science of the Total Environment 536, 432–442.

Dolman MA., Sonneveld MPW, Mollenhorst H and de Boer IJM 2014 Benchmarking the economic, Environmental and societal performance of Dutch dairy farms aiming at internal cycling of nutrients. *J Cleaner Production* 73, 245–52.

Dominati, E., Mackay, A., Green, S., Patterson, M., 2014. A soil-change based methodology for the quantification and valuation of ecosystem services from agro-ecosystems: a case study of pastoral agriculture in New Zealand. Ecol. Econ. 100, 119–129.

Dominatie, E.J., A.D. Mackay, J. Bouma and S. Green. 2016. An ecosystems approach to quantify soil performance for multiple outcomes: the future of land evaluation? Soil Sci. Soc. Amer. J. 80(2), 438–449. (doi: 10.2136/sssaj2015.07.0266).

European Commission (EC) 2006. Communication from the Commission to the Council, the European Parliament, the European Economic and Social Committee and the Committee of the Regions. Thematic Strategy for Soil Protection, COM 231 Final, Brussels.

FAO. 1976. A framework for land evaluation. FAO Soils Bulletin No. 32. FAO, Rome, Italy.

FAO. 2007. Land evaluation: Towards a revised framework Land & Water Discussion Paper 6. FAO, Rome, Italy.

FAO/IIASA/ISRIC/ISSCAS/JRC 2012. *Harmonized World Soil Database (version 1.2)*. Food and Agriculture Organization of the United Nations (FAO), International Institute for Applied Systems Analysis (IIASA), ISRIC—World Soil Information, Institute of Soil Science—Chinese Academy of Sciences (ISSCAS), Joint Research Centre of the European Commission (JRC), Laxenburg, Austria. http://www.iiasa.ac.at/Research/LUC/External-World-soil-database/HWSD_Documentation.pdf.

Field, D.J., C.L.S. Morgan & A.B. McBratney (Eds). 2017. Global Soil Security. Progress in Soil Science. Springer Int. Publisher Switzerland.

Hengl, T., J. Mendes de Jesus, R.A. McMillan, N.H. Batjes, G.B.M. Heuvelink, E. Ribeiro, et al. 2014. Soil grids 1km: Global soil information based on automated mapping. PLoS One 9(8): e105992. (12): e114788. doi: 10.1371/journal.pone.0105992.

IUSS Working Group WRB. 2015. World reference base for soil resources 2014: International soil classification system for naming soils and creating legends for soil maps. World Soil Resour. Rep. 106. Update 2015. FAO, Rome. http://www.fao.org/3/a-i3794e.pdf.

Keesstra, S.D., J. Bouma, J. Wallinga, P. Tittonell, P. Smith, A. Cerda, L. Montanarella, J. Quinton, Y. Pachepsky, W.H. van der Putten, R.D. Bardgett, S. Moolenaar, G. Mol and L.O. Fresco. 2016. The significance of soils and soil science towards realization of the United Nations Sustainable Development Goals. SOIL 2, 111–128, doi: 10.5194/soil-2-111-2016.

McBratney, A.B., J. Bouma, B.M. Whelan and T. Ancev. 2005. Future Directions of Precision Agriculture. Precision Agriculture 6 (1): 1–17.

McBratney, A.B., B. Minasny, R.A. MacMillan, and F. Carré. 2011. Digital soil mapping. In: P.M. Huang et al., editors, Handbook of soil science: Properties and processes. 2nd ed. CRC Press, Boca Raton, FL. p. 37-1-37-45.

Mc Bratney, A.B., D. Arrouays and L.A. Jarret. 2017. Quantifying Capability: Global Soil Map. In: D.J. Field, C.L.S. Morgan and A.B. McBratney (Eds). Global Soil Security. Springer Publ, 77–85.

Sanchez, P.A., S. Ahamed, F. Carré, A.E. Hartemink, J. Hempel, J. Huising, et al. 2009. Digital soil map of the world. Science 325:680–681. doi: 10.1126/science.1175084.

Soil Survey Staff. 2010. Keys to Soil Taxonomy. 11th ed. U.S. Gov. Print. Office, Washington, DC.

Stoorvogel, J.J., L. Kooistra and J. Bouma. 2015. Managing soil variability at different spatial scales as a basis for precision agriculture. Chapter 2 in: Lal, R., and Stewart, B.A. (Ed). Soil Specific Farming: Precision Agriculture. Advances in Soil Science, 37–73. CRC Press. Taylor Francis Group. Boca Raton, FL, USA.

Stoorvogel, J.J., Bakkenes, M., Temme, A.J.A.M., Batjes, N.H., and ten Brink, B.J.E. (2017a) S-World: A Global Soil Map for Environmental Modelling. Land Degrad. Develop. 28: 22–33. doi: 10.1002/ldr.2656.

Stoorvogel, J.J., Bakkenes, M., ten Brink, B.J.E., and Temme, A.J.A.M. (2017b) To what extent did we change our soils? A global comparison of natural and current conditions. Land Degrad. Develop, doi: 10.1002/ldr.2721.

Vereecken, H., J. Diels, J. Van Orshoven, J. Feyen, and J. Bouma. 1992. Functional evaluation of pedotransfer functions for the estimation of soil hydraulic properties. Soil Sci. Soc. Am. J. 56:1371–1379. doi: 10.2136/sssaj1992.03615995005600050007x.

The concept of soil security

Alex B. McBratney, Melissa Moyce, Damien Field & Alisa Bryce
Sydney Institute of Agriculture and School of Life and Environmental Sciences, The University of Sydney, Sydney, Australia

ABSTRACT: Soil security aims to maintain and improve the world's soil resource. Soil is not just a biophysical product—how we value and relate to soil affects its ability to produce the resources we need. The concept of soil security can be used to a) communicate the importance of soil science to the wider community, and b) to provide a practical route to address sustainability issues. Many sustainability goals cannot be achieved without considering the role soil has in natural ecosystems. When communicating, focusing on soil's role in these issues is more effective than focusing on soil itself. To effectively use the concept of soil security, each dimension—capability, condition, capital, connectivity, codification—must be addressed at a farm, catchment, regional, country and global scale. This method will reveal which aspect of soil security is most threatened, and provide a pathway to develop practical activities that address sustainability issues.

1 INTRODUCTION

The world that secures its soil will sustain itself (Koch et al. 2013).

Humanity faces six environmental existential challenges: food security, water security, energy sustainability, human health, climate change and biodiversity protection (McBratney et al., 2017a). Managing them in the short term is essential to accommodate the expected global population of 9+ billion and the associated pressures on natural resources, within the realm of acceptable sustainability and living standards.

These challenges are interconnected. Addressing them requires multifaceted solutions, largely based on how we value our natural resources. Natural systems are complex webs, and understanding the importance of one entity often means placing a value on exponentially more.

The importance of water, clean air, food, and most natural resources are well recognized. Yet at the nexus of these natural systems one resource lies threatened, overlooked, irreplaceable and vital to our survival.

Soil is fundamental to life on Earth.

More than ever before human society depends on products from the soil as well as the intangible services it provides. Although often hidden under the banner of agriculture, soil also plays a crucial role in climate regulation, ecosystem services, biodiversity, food security, human health, and water filtration. Soils are a key enabling resource, central to the creation of a host of goods and services integral to ecosystems and human well-being (FAO, 2015).

Sustainability is not a new concept. We recognize that as a species we are putting increasing pressure on a limited natural environment. Yet despite its crucial role in many—if not all—of the systems we depend on for survival, the importance of the soil resource is largely neglected in discussions of sustainability and more importantly, in sustainable policy-making.

To address this, McBratney et al. (2014) developed the concept of soil security. A framework than encompasses the biophysical and socio-economic aspects of soil, the concept of soil security can be used:

a. In a practical way to communicate the importance of soil science to the wider community.
b. To provide a practical route to address sustainability issues. By quantifying each dimension of soil security from the farm to global level, the greatest threats to soil and sustainability emerge.

2 SOIL SECURITY

Soil security is defined as the maintenance and improvement of the world's soil resource to produce food, fiber and fresh water, contribute to energy and climate sustainability and maintain the biodiversity and the overall protection of the ecosystem (Koch et al., 2013).

Traditional concepts of soil such as 'soil quality', 'soil health', and 'soil condition'- while useful—are narrow, vague, and generally biophysical

(McBratney et al., 2017b). Soil, like any natural resource, is more robust that the physical product itself. The value of a tree is more than the timber; it is habitat, oxygen producing, aesthetic, and more. To value soil like we value a forest, the socio-economic aspects of soil must be considered.

Security in this sense is therefore used in the same way as food, water and energy security—encompassing the economic, social and policy aspects as well as the biophysical.

The key aim in securing soil is to maintain and optimize its functionality: its structure and form, its diverse and complex ecosystems of soil biota, its nutrient cycling capacity, its roles as a substrate for growing plants, as a regulator, filter and holder of fresh water, and as a potential mediator of climate change through the sequestration of atmospheric carbon dioxide (Koch et al., 2013).

To protect a resource we must have some form of measurement of its security—some sort of value which we can use to compare and consider. Assessing and valuing soil has many aspects. Crop yield, water retention and carbon stores are just some of the obvious ways soil can be valued. But what about sources of new medicines? Or filtering toxins? What is more important—food production or conserving a threatened species on the same patch of land?

These questions highlight the complexity of assessing a natural resource. Assessing and managing soil must not only be scientific, but also contextual and value driven (Alroe & Kristensen, 2002; Bouma et al., 2012; Field, 2017; Schjonning et al., 2004). The existential challenges have complex economic, social and policy aspects which need to be addressed simultaneously (McBratney et al., 2014). Addressing soil security therefore requires a multidisciplinary and multidimensional approach (Field, 2017).

McBratney et al. (2014) suggest dividing soil security into different dimensions to distinguish between the assessment of the optimal state of the soil, the current state of the soil and how the soil is used. The resulting five dimensions are capability, condition, capital, connectivity and codification.

Capability and condition refer to the biophysical and more 'traditional' aspects of soil security. These aspects are better addressed in the literature as most existing soil science research is biophysical.

Capital addresses the socio-economic concerns which arise from soil insecurity. External perceptions of soil are captured by connectivity and codification (Kralisnikov et al., 2017). Connectivity describes the social dimension—how people, not just within the soil science community but globally, view and value the vital resource. Codification relates to the public policy and regulation necessary to achieve soil security.

2.1 Capability

Capability asks what can this soil do? (Field, 2017). This dimension recognizes that as a consequence of their intrinsic biophysical characteristics, different types of soil have different potential uses, and are influenced by the discipline of land evaluation. By measuring capability, we seek to establish what functions a particular soil can perform (McBratney et al., 2017a). Knowing the capability of a soil enables us to determine if the soil is at its full potential (a reference state) and if not, what is the potential to improve the soil in terms of the soil type, location and associated costs (Field, 2017). In other words, capability emphasizes functionality.

Capability provides the basis to quantify the resource across space and time that can be used for mapping, planning modelling and forecasting (Field & Sanderson, 2017).

Capability and condition address the traditional biophysical aspects of soil science, while connectivity, capital and codification recognise the importance of policy, social sciences and valuing soil. For modern society to value soil a number must be placed on it.

2.2 Condition

Here, condition relates to developing earlier concepts of soil quality, soil health and soil protection (Andrews et al., 2004; Karlen et al., 2001; Doran & Ziess, 2000).

When assessing condition, indicators are generally evaluated against a reference condition i.e. the capability (McBratney et al., 2017b). Soil condition can be quantified in a variety of ways. Soil organic carbon (soil carbon) is one of the most significant universal indicators of soil condition (Koch et al., 2013). Other indicators include pH, erosion, acidity, soil structure, salinity, and a variety of other biophysical parameters.

Within the concept of soil security and its five dimensions, a soil's condition must be considered in a broader sense, which is clearly defined and which allows us to measure and quantify the degree to which soil is being valued and cared for (McBratney et al., 2017b).

Condition and capability are closely related. Where capability asks what can this soil do? Condition asks can the soil continue to do this? (Field, 2017). The condition of a particular soil depends on its inherent capabilities, as well as its intended land use and the relevant management goals (Andrews et al., 2004). If soil management practices are consistent with a soil's capability, then its condition will be "fit for purpose". In other words, the use to which a soil is put should match its capability (McBratney et al., 2014).

2.3 Capital

Placing monetary values on natural resources allows us to better understand their significance. Soil is part of natural capital, defined as "the stock of materials or information contained within an ecosystem" (Costanza, 1997). Soil stocks include soil moisture, temperature, structure and organic and inorganic substances (McBratney et al., 2017a). The condition of these stocks affects the ability of the soil to provide functions known as ecosystem services.

Soil provides physical products, known as ecosystem goods (McBratney et al., 2017a). Soil also plays a role in the carbon market (McCarl, 2017). The threat of climate change has led many to suggest pursuing the mitigation of net greenhouse gas emissions through sequestration enhancement and emission control. For soil, this could involve changing practices to increase soil carbon, improving fertilization management, and changing land use from cropping to grasslands and forests (McCarl, 2017).

2.4 Codification

Codification acknowledges the need for, and role of, government policy and regulation in ensuring that soil is cared for (McBratney et al., 2017a). Effective policymaking involves the participation of all stakeholders and effective communication and translation of soil science knowledge to generate practical solutions (McBratney et al., 2017).

While the science around the connections between water, soil conditions, food and agriculture has made significant progress, there has not been a commensurate understanding of the public policy implications or the ways that these connections may or may not be understood by the public (Portney, 2017). Since 2010, an international soil policy community has emerged (Hill, 2017). Initiatives include the World Soil Charter (Food and Agriculture Organization, 1982), the World Atlas of Desertification (Barrow, 1992) and the World Soils Agenda (Hurni & Meyer, 2002). These initiatives focused on biophysical aspects of soil security, such as soil erosion and fertility. The United Nations system has begun to focus on soil as an issue for sustainable development, with an annual Global Soil Week.

This momentum is exceptionally important and valued, however, it is vulnerable to the demands of individual nations and their sustainability agendas (Hill, 2017), as each nation seeks to address the existential challenges.

Table 1. The relationship between the sustainable development goals, six existential crises, and soil.

SDG	Description	Relationship to existential crises
2. Zero hunger	End hunger, achieve food security and improve nutrition and promote sustainable agriculture.	Food
	*"ensure sustainable food production systems and implement resilient agricultural practices that increase productivity and production, that help maintain ecosystems, that strengthen capacity for adaptation to climate change, extreme weather, drought, flooding and other disasters and that progressively improve land and **soil** quality"*	
3. Healthy lives and well-being	Ensure healthy lives and promote well-being for all at all ages.	Human health
	*"substantially reduce the number of deaths and illnesses from hazardous chemicals and air, water and **soil** pollution and contamination"*	
6. Clean water and sanitation	Ensure availability and sustainable management of water and sanitation for all.	Water
7. Affordable and clean energy	Ensure access to affordable, reliable, sustainable, and modern energy for all.	Energy
12. Responsible consumption and production	Ensure sustainable consumption and production patterns.	Human health, water
	*"achieve the environmentally sound management of chemicals and all wastes throughout their life cycle, in accordance with agreed international frameworks, and significantly reduce their release to air, water and **soil** in order to minimize their adverse impacts on human health and the environment"*	
13. Climate action	Take urgent action to combat climate change and its impacts.	Climate
15. Life on land	Protect, restore and promote sustainable use of terrestrial ecosystems, sustainably manage forests, combat desertification and halt and reverse land degradation and halt biodiversity loss	Ecosystems
	*"combat desertification, restore degraded land and **soil**, including land affected by desertification, drought and floods, and strive to achieve a land degradation-neutral world"*	

3 SOIL SECURITY AS A COMMUNICATION TOOL

In 2015, the United Nations (UN) developed 17 sustainable development goals (SDG) with an aim to end poverty, protect the planet, and ensure prosperity for all. Each of the 17 SDGs are directly or indirectly dependent on soil resources (Hill, 2017). Seven have an obvious link to soil and the existential crises and four (SDG 2, 3, 12, 15) refer specifically to soil (Table 1).

Yet soil is not a specific goal. SDG15 addresses land degradation, but in a narrow and traditionally biophysical way. This is unsurprising as most soil science research is conducted in the biophysical realm. As such, soil—an already obscure science—has remained largely disconnected from policy and the public. Montanarella and Lobos Alva (2015) illustrate that soils have never been a specific focus of a multilateral environmental agreement (Keesstra et al., 2016). For example, despite its fragility and threatened nature, soil is not referred to in the International Union for Conservation of Nature (IUCN) Red List of Threatened Species (UCN, 2017) (Koch et al., 2013).

Currently, the only global initiative addressing soil degradation is the UN Convention to Combat Desertification (UNCCD), which made an urgent call for a globally recognized, measurable target for measuring land degradation and desertification (LDD) (McBratney et al., 2017a).

The SDGs offer a unique opportunity for soil scientists, policy makers and other disciplines. For soil scientists, linking soil to the SDGs is a communication opportunity. Focusing on how soil can contribute to the interdisciplinary nature of sustainability issues will be more effective than focusing on soil itself (Bouma, 2014; Bouma, 2018). For policy makers and institutions, many of the SDGs cannot be achieved without considering the role soil has in natural ecosystems.

Connecting soil to broader sustainability issues is underway in academia. Bouma (2015), Bouma (2018) and Keesstra et al., (2016) link soil science and the seven soil functions to the SDGs via ecosystem services. For example, soil function 1 'biomass production including agriculture and forestry' is essential for ecosystem service 1-providision of food, wood and fibre, and 2-provision of raw materials. These link to SDG2 1 and 2: end poverty and achieve food security. Framing soil science in the context of the SDGs in this way can communicate its importance to the outside world.

Highlighting soils' relevance through capability and condition (biophysical parameters) is a logical and essential starting point. But soil security as a wider concept offers an opportunity to identify the greatest threats to soil and humanity.

4 USING THE CONCEPT OF SOIL SECURITY

Soil security can be used as a practical means to translate sustainability issues into operational criteria (Bouma, 2018). Addressing each dimension individually will highlight which aspect of soil security is most threatened. The biggest threat may be a policy or extension issue, rather than a biophysical problem. Kidd et al. (2018) use this methodology in a case study from Tasmania, Australia.

Questions that could be asked include:

- Connectivity – is there a knowledge gap that prevents good soil management?
- Capability – is the pH, organic carbon level, salinity, etc. suitable for the current use?
- Condition – is the plan for the soil e.g. doubling agriculture production suitable. Can the soil do this?
- Codification – is the soil protected from destructive land use?
- Capital – has the soil been valued for its services to humanity?

As an example, the following section frames two of the sustainable development goals in the concept of soil security.

4.1 Sustainable development goal 2: Zero hunger

Soil security is homologous with food security. The challenge of food security is possibly the greatest threat to human society. From 2014–2016, hunger affected just over one in nine people in the world (FAO, IFAD & WFP 2015). Continuing population and consumption growth will mean that the global demand for food will increase for at least another 40 years (Godfray et al. 2010).

95–97% of the world's food comes from soils (FAO & ITPS 2015; McBratney et al. 2017a). Over 99% of food energy intake comes from crops grown on soil; less than 0.3% comes from aquatic sources. Effective management of our soils can lead to increased crop yields. For example, Lal (2004) found that an increase of 1 ton of soil carbon pool of degraded cropland soils may increase crop yield by 20 to 40 kilograms per hectare (kg/ha) for wheat, 10 to 20 kg/ha for maize, and 0/5 to 1 kg/ha for cowpeas.

Target 2.3 states *"By 2030, **double the agricultural productivity** and incomes of small-scale food producers, in particular women, indigenous peoples, family farmers, pastoralists and fishers, **including through secure and equal access to land**, other productive resources and inputs, knowledge, financial services, markets and opportunities for value addition and non-farm employment."*

4.2 Condition and capability

Can the soil produce the required food (capability), and can it keep doing so? (condition). Doubling the agricultural productivity of small-scale farmers requires soil in good enough condition now, and with the capability to be improved to achieve this goal.

4.3 Connectivity, capital and codification

Does the land manager have the right knowledge and resources to manage the land according to its capability? (connectivity). Connection to the soil often comes through land tenure. Katz (2000) noted that soil is used less optimally where land tenure is unclear/absent. Indeed, poor agricultural performance and associated land degradation have been blamed on a lack of land tenure policies (codification) (Burgi, 2008).

Address competing claims—is the soil valued more for its food production capabilities, ecosystem services, biodiversity, or ability to mitigate climate change (capital)? Places in the world that are most food insecure are also often regions with competing demands and existing soil degradation. Hotspots of food security endangered by land degradation are recognized in Africa and Southern Asia (Krasilnikov et al., 2017). Some of these regions host valuable biodiversity and delivering ecosystem services of global importance (Hooper et al., 2005), "which leads to competing claims between local and international communities" (Keesstra et al., 2016).

On a global scale this is a difficult problem to address. But at the farm scale, deciding which of these is the greatest threat for the situation provides a starting point to address the issue.

4.4 Sustainable development goal 15: Life on land

"There is currently an unprecedented threat to the world's soils through degradation" (McBratney et al., 2017a). Overexploitation of vegetation and soil resources, together with inappropriate agricultural systems, is resulting in accelerated rates of land degradation, soil erosion and nutrient depletion (Hill, 2017). The UNCCD initiative was developed in response to its contribution to biodiversity loss, climate change mitigation and alleviation of poverty. This falls short, however, as it only addressed arid and semiarid landscapes rather than the full range of threats to the world's soils and the services they provide (McBratney et al., 2017a).

Soil degradation inherently reduces or eliminates soil functions and their ability to support ecosystem services essential for human well-being (FAO, 2015). The degradation of soil through erosion, fertility decline, acidification, salinity, compaction and soil carbon decline has significant consequences for agricultural productivity, provision of water and loss of biodiversity (Field, 2017).

SDG 15 specifically addresses land degradation issues. Target 15.3 is "By 2030, combat desertification, restore degraded land and soil, including land affected by desertification, drought and floods, and strive to achieve a land degradation-neutral world."

This goal recognizes that soil condition and capability are declining. To achieve degradation neutrality, a soil's biophysical decline must be halted and reversed.

The causes of decline are, however, not biophysical at heart. Condition and capability are not the greatest threats. How we value, connect to and protect the soil influence the degree of degradation. Cerda and Keesstra in Keesstra et al., (2016) use the example of the Mediterranean: Changing socio-economic conditions in the last 30 years have converted agriculture from a sustainable system to an unsustainable system. The economically attractive agricultural intensification with drip irrigation and readily available nutrients have reduced the soils 'value' to the land owner. Rather than a medium that works symbiotically with the crop, soil has become merely a sponge to hold water and nutrients until required. It could be argued that this loss of connectivity with the soil, the 'mental de-valuing', has been a key driver of degradation in this situation.

SDG15 has attracted "considerable attention in policy making circles as well as civil society" (Desai and Sidhu, 2017). This goal has been codified at a global scale, but much more work is required as policy at the domestic level is necessary to practical outcomes (Desai and Sidhu, 2017).

5 WHAT'S NEXT FOR SOIL SECURITY

The concept of soil security is a way to measure and value a threatened resource. It has strategic value in that it can serve to focus and guide the development of policies addressing the six global existential challenges, such that interventions for one challenge result in favourable effects on the other (McBratney et al., 2017a).

The challenge now is to develop a system of quantifying the five dimensions at a farm, catchment, regional, country and global scale. Numerous papers in Field et al. (2017) provide clues on how to do this e.g. McBratney et al. (2017d), Wills et al. (2017), Morgan et al. (2017), Kim et al. (2017).

Soil scientists are good a quantifying soil capability, and to a lesser extent condition. However globally, the socioeconomic dimensions of soil security are poorly understood. That soil is not its own SDG indicates a severe lack of global connectivity. Ecosystem services provide some measure of value, but the process is rudimentary. Use value for production is not very well quantified, even by property values. As a result there is little codification to protect this essential and non-renewable resource.

For soil security to be achieved, we need local, national and international agreements around soil. These agreements must expand on existing biophysical soil knowledge, and address the socio-economic aspects necessary to protect this natural resource.

Human pressures on soil resources are reaching critical limits (FAO & ITPS, 2015). The pressures on soil are widespread and varied, and the challenges created by these demands deeply affect our ability to provide sufficient resources for the world's growing population (Koch et al., 2013). Soil is relevant and as essential to life on earth as water, air and light. It is time it takes its place as an equal. For without soil, there is no humanity, and no world to protect.

REFERENCES

Alrøe, H.F., Kristensen, E.S., 2002. Towards a systemic research methodology in agriculture: rethinking the role of values in science. Agric. Hum. Values 19, 3–23.

Andrews, S.S., Karlen, D.L., Cambardella, C.A. 2004. 'The Soil Management Assessment Framework: A Quantitative Soil Quality Evaluation Method' Soil Sci. Soc. Am. J. 69: 1945–1962.

Barrow, C.J. 1992 World atlas of desertification (United nations environment programme), edited by N. Middleton and D.S.G. Thomas. Edward Arnold, London.

Bouma, J., Stoorvogel, J.J., Sonneveld, W.M.P., 2012. Land evaluation for landscape units. In: Huang, P.M., Li, Y., Summer, M. (Eds.), Handbook of Soil Science. CRC Press, Boca Raton London, New York, pp. 34-1–34-22.

Bouma, J. 2014. Soil science contributions towards Sustainable Development Goals and their implementation: linking soil functions with ecosystem services. J. Plant Nutr. Soil Sci. 177, 111–120.

Bouma, J. 2015. Engaging soil science in transdisciplinary research: facing wicked problems in the information society. Soil Sci. Soc. Am. J. 79, 454–458.

Bouma, J. 2018. How the soil security concept can pave the way to realizing some soil related UN-SDG's. In "Global Soil Security—Towards More Science-Society Interfaces" (this book). Richer-de-Forges A.C., Carré F., McBratney A.B., Bouma J., Arrouays D. (Eds.). CRC Press Taylor & Francis Group.

Burgi, J.T. 2008. 'The dynamics of tenure security, agricultural production and environmental degradation in Africa: evidence from stakeholders in north-east Ghana'. Land Use Policy. 25: 271–285.

Costanza, R. 1997. 'The value of the world's ecosystem services and natural capital' Nature. 387: 253–260.

Decaëns, T., Jiménez, J.J., Gioia, C., Measey, G.J., Lavelle, P. 2006. 'The values of soil animals for conservation biology' European Journal of Soil Biology. 42(1): S23-S38.

Desai, B.H. and Sidhu, B.K. 2017. Striving for Land-Soil Sustainability: some legal reflections. In 'International Yearbook of Soil Law and Policy', Springer. Ginzky et al (Ed): 37–45.

Doran, J.W., Ziess, M.R., 2000. 'Soil health and sustainability: managing the biotic component of soil quality' Appl. Soil Ecol. 15, 3–11.

FAO 2015. World Soil Charter Available at: http://www.fao.org/fileadmin/user_upload/GSP/docs/ITPS_Pillars/annexVII_WSC.pdf, Accessed 21 June 2017.

FAO, IFAD and WFP 2015. The State of Food Insecurity in the World 2015. Meeting the 2015 International hunger targets: taking stock of uneven progress. Rome, FAO.

FAO and ITPS 2015. 'Status of the World's Soil Resources (SWSR): Main Report' Food and Agriculture Organization of the United Nations and Intergovernmental Technical Panel on Soils, Rome, Italy.

Field, D.J. 2017. 'Soil Security: Dimensions' Global Soil Security: 15–23.

Field, D.J. & Sanderson T. 2017. 'Distinguishing Between Capability and Condition' Global Soil Security: 45–52.

Field, D.J., Morgan, C.L. & McBratney, A.B (Eds.). 2017. Global Soil Security. Springer

Food and Agriculture Organization, 1982. World Soil Charter. FAO, United Nations Publication.

Godfray, H.C.J., Beddington, J.R., Crute, I.R., Haddad, L., Lawrence, D., Muir, J.F., Pretty, J., Robinson, S., Thomas, S.M., Toulmin, C. 2010. 'Food Security: The Challenge of Feeding 9 Billion People' Science. 327(812): 812–818.

Hill, R. 2017. 'The Place of Soil in International Government Policy' Global Soil Security: 443–449.

Hooper, D.U., Chapin, IiiF., Ewel, J., Hector, A., Inchausti, P., Lavorel, S., Lawton, J., Lodge, D., Loreau, M. and Naeem, S. 2005. 'Effects of biodiversity on ecosystem functioning: a consensus of current knowledge', Ecol. Monographs. 75: 3–35.

Hurni, H., Meyer, K. (Eds.), 2002. A World Soils Agenda, Discussing International Actions for the Sustainable Use of Soils, IASUS Working Group of the International Union of Soil Sciences (IUSS). Centre for Development and Environment, Berne.

Karlen, D.L., Andrews, S.S., Doran, J.W., 2001. Soil quality: current concepts and applications. Adv. Agron. 74, 1–39.

Katz, E.G. 2000. 'Social capital and natural capital: a comparative analysis of land tenure and natural resource management in Guatemala'. Land Econ. 76: 114–132.

Kidd, D., Field, D., McBratney, A & Webb, M 2018. A preliminary quantification of the soil security dimensions for Tasmania, Geoderma. 322: 184–200.

Kim, S.C., Lim, K.J., Yang, J.E. 2017. 'The Measurement of Soil Security in Terms of Human Health: Examples and Ideas' Global Soil Security: 297–304.

Koch, A., McBratney, A., Adams, M., Field, D., Hill, R., Crawford, J., Minasny, B., Lal, R., Abbott,

L., O'Donnell, A., Angers, D., Baldock, J., Barbier, E., Binkley, D., Parton, W., Wall, D.H., Bird, M., Bouma, J., Chenu, C., Flora, C.B., Goulding, K., Grunwald, S., Hempel, J., Jastrow, J., Lehmann, J., Lorenz, K., Morgan, C.L., Rice, C.W., Whitehead, D., Young, I., Zimmermann, M., 2013. 'Soil Security: Solving the Global Soil Crisis' *Global Policy*. 4(4): 434–441.

Krasilnikov, P., Sorokin, A., Mirzabaev, M., Makarov, Strokov, A. and Kiselev, S. 2017. 'Economics of Land Degradation to Estimate Capital Value of Soil in Eurasia' *Global Soil Security*: 237–246.

Keesstra, S.D., Bouma, J., Wallinga, J., Tittonell, P., Smith, P., Cerdà, A., Montanarella, L., Quinton, J.N., Pachepsky, Y., van der Putten, W.H., Bardgett, R.D., Moolenaar, S., Mol, G., Jansen, B and Fresco, L.O. 2016. 'The significance of soils and soil science towards realization of the United Nations Sustainable Development Goals' *SOIL*. 2:111–128.

Lal, R. 2004. 'Soil Carbon Sequestration Impacts on Global Climate Change and Food Security' Science.34(5677): 1623–1627, American Association for the Advancement of Science.

McBratney, A.B., Field, D.J., Koch, A. 2014. 'The dimensions of soil security' *Geoderma*. 213: 203–213.

McBratney, A.B., Field, D.J., Morgan, C.L.S., Jarrett, L.E. 2017a. 'Soil Security: A Rationale' *Global Soil Security*: 3–14.

McBratney, A.B., Field, D.J., Morgan, C.L.S., Jarrett, L.E. 2017b. 'General Concepts of Valuing and Caring for Soil' *Global Soil Security*: 101–108.

McBratney, A.B., Field, D.J., Morgan, C.L.S., Jarrett, L.E. 2017c. 'The Value of Soil's Contribution to Ecosystem Services' *Global Soil Security*: 227–235.

McBratney, A.B., Field, D.J., Arrouays, D., Jarrett, L.E. 2017d. 'Quantifying Capability: GLobalSoilMap' *Global Soil Security*: 77–86.

McCarl, B.A. 2017. 'Economics, Energy, Climate Change, and Soil Security' *Global Soil Security*: 195–205.

Montanarella, L. & Lobos Alva, I. 2015. 'Putting soils on the agenda: The Three Rio Conventions and the post-2015 Development Agenda' *Curr. Opin. Environ. Sustain.*, 15, 41–48.

Morgan, L.S., Morgan, G.D., Bagnall, D. 2017. 'Soil Licensing to Secure Soil' *Global Soil Security*: 247–254.

Portney, K.E. 2017. 'Soil-Water-Food Nexus: A Public Opinion and Policy Perspective' *Global Soil Security*: 371–381.

Schjønning, P., Elmholt, S., Christensen, B.T., 2004. Soil quality management—concepts and terms. In: Schjønning, P., Elmholt, S., Christensen, B.T. (Eds.), *Managing Soil Quality: Challenges in Modern Agriculture*. CABI Publishing, UK, pp. 1–17.

UCN (2017) The IUCN Red List of Threatened Species. Version 2017.1 [online]. Available from: http://www.iucnredlist.org [Accessed 11 June 2017].

Wills, S., Williams, C., Seybold, C., Scheffe, L., Libohova, Z., Hoover, D., Talbt, C., Brown J. 2017. 'Using Soil Survey to Assess and Predict Soil Condition and Change' *Global Soil Security*: 123–136.

Approach to valuing soil ecosystem services with linking indicators

Dianna K. Bagnall, Cristine L.S. Morgan, Richard T. Woodward & Wm. Alex McIntosh
Texas A&M University, College Station, Texas, USA

ABSTRACT: Valuing the contribution of soil to human welfare is a critical step in securing the global soil resource. Secure soil contributes to human wellbeing through soil Ecosystem Services (ES). Cooperation between soil scientists and social scientists is needed to convey value to decision makers. Linking indicators are proposed as a tool for conveying this value. These indicators will be audience specific and will be chosen considering the context of each Soil Security dimension. Use of linking indicators is particularly recommended for the contingent choice, benefits transfer, and damage cost avoidance methods of valuation. The benefit transfer method requires knowledge of soil heterogeneity at spatial scales that soil scientists can provide. The productivity method of valuation will prove to be a powerful tool to determine the contribution of soil ES in monetary units, but requires attention to the functions underlying partitioning of ES benefits between soil and other contributors.

1 INTRODUCTION

Soil Security is a concept recently developed by soil scientists for the purpose of framing soil science research and outreach in relevant social, political, and economic contexts. Of the five dimensions of Soil Security (McBratney, Field, and Koch, 2013), the two that have traditionally been focused on by soil scientists, capability and condition, are mature and deployable relative to the remaining three dimensions. The codification dimension focuses on policy and requires maturity in the other four dimensions. The remaining two dimensions, connectivity and capital, have been the focus of this work because they are the weakest links to securing the global soil resource.

The capital dimension of Soil Security focuses on valuing ecosystems services (ES) that soil provides to people and communicating that value to decision makers. Valuation in this context should not be considered a synonym for monetization—biophysical metrics can convey value extremely well when presented to the appropriate audience. Converting these metrics to monetary units may add uncertainty in some cases and/or complicate communication.

Soil scientists have used valuations such as damage associated with erosion (Smith, 2014; Duffy, 2012) or partial budgets for new technologies (Townsend et al., 2016) to convince land managers to change practices. The limited success of these methods may stem from the need for communication methods that are credible, salient, and legitimate (Ingram et al, 2016). Additionally, soil scientists must recognize that the many distinct audiences that they need to communicate with will require different metrics and language (Reimer et al, 2014). The goal of this paper is to address the problems of valuing the soil resource in such a way that values can be communicated to stakeholders and decision makers.

2 VALUING SOIL

The capital dimension of Soil Security seeks to value soil based on the understanding that "things" that are not valued will not be conserved (Robinson et al., 2009). This realization in soil science mirrors valuation attempts by other biophysical sciences as the transdiscipline of environmental economics emerged. One of the tasks of environmental economics has been to value ES, which are the benefits people obtain from ecosystems (MEA, 2005). The ES concept is anthropocentric because of its focuses on human welfare. It also recognizes that market signals do not provide appropriate information for management of ecosystems. This lack of information can be filled by incorporating ES into markets or by providing non-market values for decision makers to consider.

Soil has been considered in markets as a raw material (Robinson et al, 2009). Valuing raw materials or products that come from soil, even if that product requires extraction of the soil itself, does not allow us to know what value that soil would provide if left in place—only that the private value in place is less than the value paid for the soil when extracted. Poor soil conservation practices may not cause a reduction in the market price of a product or material, but they may reduce provision of other soil (ES). The question is, how are the values of these other soil ES experienced by the person who makes the land management decision? Under current paradigms, they may not be experienced at all.

A major focus of universities and soil societies has been to educate land managers about how soil management practices provide monetary benefits (e.g. Smith, 2014; Duffy, 2012). But neither pricing as a raw material nor valuing the goods that soil contributes are adequate valuations of soil (McBratney, 2013; Koch, 2013; Knight, Robins, and Chan, 2013).

Operationalizing Soil Security requires that the value of all soil ES be experienced by the person who makes the decision whether to conserve soil or not, so estimates of soil ES value that are not (or not yet) captured by market price are needed.

The ES concepts hold promise as a means for soil valuation. Since at least the 1960s, valuation estimates have been made for many ES (deGroot et al., 2012). In their textbook on ecological economics, Daly and Farley (2011) list and describe the following eight valuation methods:

1. Market Price Method
2. Productivity Method
3. Hedonic Pricing Method
4. Travel Cost Method
5. Damages Cost Avoidance Method
6. Contingent Valuation (stated preference) Method
7. Contingent Choice Method
8. Benefit Transfer Method

In situations in which soil is a market good (e.g.: sale of top soil), assigning a value is simple—it is the price people are willing to pay (soil is not renewable in human timescales and it typically not considered an ES for this reason). However, the soil would have provided other benefits that do not have market prices if it had been left in place and those benefits may not be accounted for in its market price. Benefits of the soil that are public in nature, such as its role in providing habitat to migratory birds, will not be reflected in market prices. This simple example illustrates that any effort to value soil must capture both market and non-market values. While no singular valuation method will capture all the values of soil, changes in soil function, and therefore changes in provision of welfare to humans, can be quantified as aids to benefit/cost analysis and decision making. In seeking to value soil ES, soil scientists can learn lessons from previous attempts to value other ES.

Perhaps the most extreme example, and one that has been widely criticized in economic circles, was the effort to capture the value of global ES by Costanza et al. (1997). In this case the benefit transfer method was used to extend previous contingent valuations of ES at local and regional scales to a global scale. Essentially, ranges of ES values for given biomes were multiplied by the area of those biomes across the globe. The estimated annual value of global ES in 1934 US dollars was $33 trillion. Costanza noted that these values were largely not captured by markets. The work was updated in 2014 to reflect current dollar figures and include more studies. This work has drawn attention and criticism to the valuation of ES and the methods used. A particular concern is the global application of values determined at local or regional scales (Konarska, 2002; Cohen, 2006). It has also been noted that the indictors used were not linked to values experienced by particular individuals in space and time (Toman, 1998).

While imperfect, Costanza et al. (1997) and similar studies (de Groot et al., 2012) have been a starting point for recent efforts of soil scientists to value the global soil resource. As a contribution to the 2015 Global Soil Security Conference, McBratney et al. (2016) estimated the total annual value of the soil contribution to the ES that Costanza listed. The methodology was to determine what fraction of the Costanza (1997) value was contributed by soil. The total annual value of the soil calculated, $11,831 billion, is almost seven times larger than the world GDP. Both Costanza (1994) and McBratney (2015) discuss the reasons that their values should be considered as minimums and note that ES will likely become scarcer, and more valuable, in the future. However, McBratney et al.'s value is subject to the same criticisms as Costanza's 1997 work, namely that it is not experienced by a particular person in time and space. More work is needed to reinforce such average global values with those that are experienced by particular people. Toman's critique of Costanza et al. (1997) concluded with a call for interdisciplinary research both on the assessment of specific values and the use of those values—a call that has broad support in both social and biophysical sciences (Bouma, 2016; Boyd et al., 2014; Koch et al., 2013; Knight, Robins, and Chan, 2013; Robinson et al., 2009; Bouma, 2001).

3 LINKING INDICATORS

Interdisciplinary research on the assessment and use of soil values begins with developing a common language. Boyd et al. (2014) suggests that a major limitation for interdisciplinary research is the lack of common indicators between disciplines. The needed pieces of information are termed "linking indicators" and are described as a subset of biophysical indicators that best facilitate social interpretation of ecological conditions and change (Boyd, et al. 2014). Linking indicators may be measurements or model outputs, or they may be calculated or arrived at from measurements or model outputs. Recognizing that not all valuable soil ES are accounted for in market transactions, it is critical that decision makers have supplementary information, aside from market prices, that inform their decisions. Selecting appropriate biophysical metrics for use as linking indicators can provide this information.

Linking indicators are expected to be stakeholder specific, (Johnston and Russell, 2011; Reimer et al. 2014) which means that useful indicators will convey value only when presented to the appropriate audience(s). They should not be so complex that an audience requires training to understand them, but neither should they be overly simplified—it is not intended that all data be presented at the level of the least informed audience.

Let us consider an example in which a soil scientist measures changes in soil function resulting from a farmer's changes in soil management. If the soil scientist measures the increased water infiltration and storage resulting from changes in soil management, the available water for crops can be reported with confidence. If the soil scientist takes that volume and wants to convert it to a monetary unit, other factors must be considered, such as the farmer's access to irritation, the cost of that irrigation, the rainfall in that particular year (which may make additional water either very valuable or potentially damaging), the crop grown, and the price of the crop for that year. The forgoing list of factors are understood by the farmer, making her the most qualified person to assess the value of the additional volume of water. On the other hand, an outside analyst, be it a soil scientist or an economist, can only estimate the value of that change in water availability; and while that estimate might be right on average, it may be far off for any given farmer. Hence, in terms of providing a useful indicator of the benefits of soil management, farmers are likely to see a measure of change in the capacity of the soil to hold water as more informative than a simple dollar value.

Linking indicators should reduce the audience's need to speculate about biophysical relationships. The above example illustrates that the linking indictor is preferable to an estimated cost in this case, but is not intended to imply that direct measurements should always be reported to the audience. Using the same example, the soil scientist could have reported increases in infiltration or in volumetric water content rather than an additional capacity of the soil to hold water. While there is less uncertainty in these indicators, they are unlikely to be linking indicators because the farmer is unlikely to clearly and directly associate value with those measurements. The famer may know that more infiltration results in more water for crops, but without knowing the nature of that relationship, the farmer is not able to evaluate if there is a significant value of water resulting from the change in soil management practices. In this example, the linking indicator was the additional water holding capacity of the soil, which was calculated based on increased infiltration and storage of water in soil as the result of a soil management change. The volume of water is a better linking indicator than the measurements of storage and infiltration; it clearly and directly mattered to the farmer. The linking indicator is also better than monetary values because to obtain monetary units requires assumptions that may not be correct for any given farmer.

In the example given, the soil manager was also the person who experienced the value of a soil ES. When the soil manager does not experience the cost or benefit of a decision that impacts soil ES (an externality in economic terms), selection of indicators is more complex. Consider a farmer whose land drains into a river. High levels of soil erosion may be economical for the farmer either because conservation is expensive or the benefits of control are slight. But this high erosion may be harming recreational fishing. Measurements of erosion are not linking indicators for the farmer or for the recreational angler. The measurements of desirable fish population are likely to be a linking indicator for the angler, but not the farmer, who makes the decision about erosion management practices. However, because the angler can value what a fish population is worth to him or her, this linking indicator provides valuable information that anglers can use to attempt to convince the farmer to reduce erosion, perhaps by paying the farmer for ES. Regulatory and community based options for such negotiations are possible, but beyond the scope of this paper.

A global map of soil value in monetary units is attractive because it would seem to appeal to everyone. In reality, the values on such a map might not be experienced by anyone. A given volume of nitrogen gained by planting a cover crop has real and experienced value to a farmer and reductions in parts per millions of a contaminant entering a water treatment plant has real value to the plant manager. But these indicators are audience specific. This audience specificity means that using biophysical metrics as linking indicators will be more cumbersome for large national or global assessments, but they should provide stronger and more pertinent evidence to soil managers and local or regional decision makers. Additionally, soil scientists have been addressing problems of scale and heterogeneity with mapping, pedotransfer functions (functions that use regression analysis and data mining techniques to extract rules associating basic soil properties with more difficult to measure properties), and soil classification for the history of the discipline. These tools enable soil scientists to meaning fully map linking indictors across space.

In detailing lessons learned from using the InVEST Model to convey the value of ES to decision makers, Ruckelshaus et al. (2015) noted that often policy and decision makers preferred biophysical units to monetary units. In many cases, the decision makers wanted a mix of monetary and biophysical values that represented the outcomes of choices. This observation supports the idea that the use of best indicators will be dependent on audiences and on situations.

4 APPLYING VALUATION METHODS AND LINKING INDICATORS TO SOIL ECOSYSTEM SERVICES

Dominati et al. (2010a, 2014b) list a subset of ES to which soils have critical contributions and these serve as a practical starting point for valuation. Linking indicators are useful when valuation methods seek to describe costs and benefits in the absence of, or as complements to, a price. Each soil ES can be linked with one or more valuation methods as illustrated in Figure 1.

When discussing valuation, one should keep in mind that it should be used to assess changes in conditions, rather than the total value of a state. For example, it would be reasonable to estimates the value of an improvement in a soil's condition, which might lead to more yields, less erosion, and so on. The total value, on the other hand is usually much less meaningful. The total value of a soil would be the difference between the value that the soil generates and the value of the situation in which the soil is completely gone, holding all else equal. But of course, if there is no soil left, very little else can be equal—the situations are so radically different that they are essentially non-comparable. Hence, the methods discussed below are best applied to changes in soil condition.

4.1 Contingent valuation and stated choice methods

Contingent valuation methods try to determine the value of a good or service based on what individuals directly say they would be willing to pay for it. A contingent choice study tries to arrive at the value of a good or service by asking how much of a similar good or service an individual would

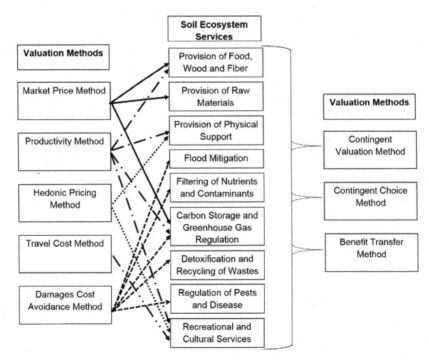

Figure 1. Compatibility of valuation methods and soil ES. Valuation methods that are appropriate for particular soil ES are linked with an arrow.

be willing to substitute. These methods have been criticized (Hausman, 2012), but have been used by industry and government in notable cases like the British Petroleum oil spill (Kling, Phaneuf, and Zhao, 2012) and have been defended in literature (Haab et al., 2013). Soil scientists should collaborate with social scientists early in project planning to find appropriate ways to apply contingent value and contingent choice methods.

Contingent valuation studies almost always generate monetary values, since respondents are asked about their willingness to pay. Linking indicators still can play a role in such studies, however as they can be an expression of the services that respondents might be willing to pay for (e.g., soil functions or clean water).

4.2 Market price method

Market Price valuation is perhaps the simplest form of soil ES valuation—it estimates values for soil ecosystem products or services that are bought and sold in commercial markets. This method should be used for products that are produced on soil such as food, fiber, and lumber, as well as for raw materials extracted from soils. Soil that is mined as a raw material is not classified as an ES because it is functionally non-renewable (soil formation is closer to a geologic than human time-scale.) Many on-farm economic studies could be called market price studies. A factor (management or input) is changed and the resulting change in market benefits is measured. If the increase in market benefits is greater than the cost, the change in the factor is considered to have positive economic value. The concept of additional value is key in these valuations. Crops are not often grown without soil, but the presence of soil, even excellent quality soil, will not produce that crop. If altered management can be solidly linked to a service that increases the market value, then we can use the market price method for the soil ES valuation. Carbon storage and greenhouse gas regulation can be included as soil ES that can be valued with market pricing in any context in which the soil manager can experience those values (ex: in a carbon market).

4.3 Productivity method

The productivity method, also called the derived value method, can be used to estimate the economic value of soil ES that contribute to the production of commercially marketed goods. This method can be used in the provision of food, wood, and fiber, provision of physical support, carbon capture and greenhouse gas regulation, and recreational and cultural services. An example of the productivity method for valuing the provision of food, wood, or fiber is a link between soil infiltration rate and water captured. The captured water can be experienced as reduced irrigation costs or in increased yield from a crop. Soil infiltration rates cannot be marketed directly, but the water captured because of those increased infiltration rates can be quantified. This captured water has value to a farmer because it can be translated to money saved on irrigation costs or money earned through increased crop yield.

A productivity model could estimate the value contributed by soils to physical support that is valued for biomass production or for human infrastructure. Some soils, for example a shallow soil over bedrock, may contribute minimally to physical support of infrastructure, while others, such as deep shrink-swell clays, may play critical roles in what infrastructure can be developed on them. Apart from the direct pricing of carbon storage and greenhouse gas regulation discussed above, there are situations in which soil contributes in a less direct manner to this ES. Even when no direct market is available for carbon, carbon storage may still be valuable to farmers if it contributes to marketable goods that can be valued. The productivity method is linked to recreational and cultural services in that soil enables many of these services even though it does not wholly provide them. Examples are quality of recreational areas, quantity and quality of recreational species (eg: trout habitat related to erosion), and burial sites.

However, it is critical to note that the productivity method is only as accurate or precise as the functions that are used to partition values between soil and some other contributor to ES. For example, we could arbitrarily claim that soil contributes 50% of the value to recreational and cultural services that take place on soil and multiply all values of such services by 0.5 to estimate the soil ES value of recreational and cultural services. This valuation is unhelpful because it is difficult to validate and is not likely to be experienced by decision makers. Valuation by the productivity method must be based on clear social, biophysical, and economic relationships.

4.4 Hedonic pricing method

Hedonic pricing as a means of valuation is most commonly applied to variations in housing or land prices that are linked to environmental attributes. To whatever extent soil ES can be valued as recreational and cultural services, they may translate to hedonic prices. Degradation of soil, such as pollution, erosion, or a known soil issue such as propensity for mudslides, may impact land prices depending on how publicized a pollution issue is and the intended land use. Hedonic pricing could be used in cases where the structural benefits provided by soils add or remove value to homes or when a soil's capability limits development (buildings on shrink-swell clays may be more expensive, drainage for septic systems may be restricted by low permeability soils). Capability and condition of soil are major contributors to the productivity of land. How this translates to increased or decreased values of land is variable. The yield average for a parcel of crop land can be used as a proxy for the productivity of that land. Identifying how those yields are affected by soil capability and condition is complex. Research is needed to evaluate land sales that have ranges in soil capability and condition to determine to what degree pricing of land considers soil capability and condition. In the absence of hedonic prices, linking indicators could be used to compare properties based on their biophysical attributes, but such comparisons would not be hedonic pricing *per se*.

4.5 Travel cost method

The travel cost method uses the expenses that people incur to reach and experience a location as a proxy for the price to use that location, and that can then be used to estimate the demand for the location in a way similar to the market price approach. This is unlikely to be a useful method for the direct estimate of a soil's value unless and until rare and endangered soils are visited. With a more solid understanding of soil's contribution to cultural and recreational services, the travel cost method should be able to extend to soil valuation. It could be possible to use some fraction of travel cost along with the productivity method to estimate the value of the contribution of soil ES to all locations that people travel to. That is, if the location had a massive loss of soil and became very unsightly, people would not to travel to it. Short of this catastrophe, travel to that location is not likely to be impacted. Situations in which a tremendous amount of loss of soil ES result in reduced travel are better suited to the damage cost avoidance method of valuation. Linking indicators are unlikely to be used for the travel cost method of valuation *per se*, but could be very useful in assessments of tourist destinations. Ruckelshaus et al. (2015) notes that environmental managers who have tourism in mind are often interested in key biophysical indicators of tourism for their sites, such as fish population or wildlife diversity. Linking indictors for soil ES may play a role in aggregate assessments of the likelihood that a location will attract tourism.

4.6 Damages cost avoidance method

Costs can be determined for damages resulting from flooding, nutrients and contaminants, greenhouse gases, waste, pests, and disease. To the degree that these damages and their costs can be avoided by maintenance or improvement of a soil resource, those avoided costs represent valuable services provide by the soil. Clarity of relationships between damages and soil capability and condition will vary among soil ES. Flooding is an event that costs will be clear for—costs of replacing damaged goods or infrastructure or of constructing man-made flood mitigation infrastructure can be assessed. Greenhouse gas regulation is more complex because the costs associated with increased greenhouse gas concentrations are dispersed across the globe making it difficult to link causes and effects. The cost of filtering of nutrients and contaminants, detoxification of wastes, and regulation of pests and disease may be measured by comparison of the efficiency and cost of the soil ES to an equivalent engineered solution (filtration, detoxification, pest/disease control).

Linking indicators can play an important role in damage cost avoidance as a means for valuing soil ES. In predicting the events that will cause damages, there is likely greater certainty in the biophysical predictions than to the costs. Predicting a flood will first require our prediction of biophysical metrics (e.g. the hydrograph) before a cost is assigned. Also, similar to our example of uncertainty of factors affecting farm income, there are many factors that complicate the dollar value associated with damages from biophysical events. Providing metrics that managers and planners can clearly value in terms of damage cost avoidance is a promising application of linking indicators for valuation of soil ES.

4.7 Benefit transfer method

The benefit transfer method estimates the value at one place using values from other studies and locations. Whether the benefit transfer method is appropriate for a soil ES valuation depends on whether the locations and contexts to which benefits are being transferred are sufficiently similar to those where the information was collected. Each dimension of soil security would need to be addressed. For capability, similar soils under similar climatic conditions may be good candidates for benefit transfer valuation. Soil maps could play key roles in determining applicability to other locations, because soil maps contain capability information (depth, particle size, etc.). However, the condition of the soils should also be considered. Examples of soil conditions are degree of soil structure, soil pH, and soil nutrient status. Soil maps are unlikely to be accurate gauges of the similarities between soil conditions. Hence, caution must be used, critically considering that the conditions of the soils that are being compared are comparable.

In addition to soil capability and condition, benefit-transfer estimates should control for capital (economic), connectivity (social), and codification (policy) contexts. The value of soil carbon that was assessed on a soil in a country that has an established carbon market should not be applied to a similar soil in a country that has no access to a market. While complex, the considerations that need to be accounted for in benefit transfer methods are similar to those needed for the use of pedotransfer functions and in much of the soil data that soil scientists wish to apply to areas larger than their test sites. Soil scientists are well equipped to determine the appropriate benefit transfers for soil capability and condition, and now need to collaborate to add the appropriate social, political, and economic contexts.

The main challenge in the selection of linking indicators for benefits transfer will be making sure that the biophysical and social settings that benefits are transferred to are sufficiently like those that benefits are being transferred from.

5 SUMMARY AND CONCLUSIONS

Soil Security is a concept recently developed by soil scientists for the purpose of framing soil science research and outreach in relevant social, policy, and economic contexts. The capital dimension of soil security seeks to value soil and can be informed by the work of environmental economics in valuing soil ES. Methods for valuing ES include: contingent value, contingent choice, market price, productivity, hedonic price, travel cost, damage cost avoidance, and benefit transfer methods. The connectivity dimension of soil security is concerned with people's connection to and awareness of soil. Cooperation between soil and social scientists is needed to design experiments that convey real and experienced value to the people who have the power to make decisions about soil use and conservation. Linking indicators are proposed as a tool for conveying the value of soil to soil managers and decision makers. These indicators will be audience specific and will need to be chosen within the context of each Soil Security dimension.

Of the eight ES valuation methods discussed, we recommend that linking indicators be used in at least three: contingent choice, benefits transfer, and damage cost avoidance. Linking indicators are nearly always an important part of assessing ES. Clear statements about linking indicators can be a critical part of good valuation, or might be an alternative way to express the outcomes of concern.

The benefit transfer method requires knowledge of soil heterogeneity at spatial scales that soil scientists can provide. Projects featuring benefit transfers of soil linking indicators must be considered in appropriate policy, social, and economic contexts. The productivity method could be a powerful tool to determine the contribution of soil ES in monetary units, but requires rigorous attention to the functions underlying the partitioning of ES benefits between soil and other contributors.

There will be situations in which the person who experiences the consequences of changes in soil function is not the person with ability or authority to make changes to soil capability or conditions— an externality in economic terms. This paper has not attempted to address the changes in social or political structure needed to internalize these externalities. This would mean that the person who can make the decision would need to experience the consequences of that decision.

The immediate need for Soil Security is for projects grounded in the principles of all five dimensions to be deployed and act as case studies for conveying experienced soil values to soil managers and decision makers for the purpose of soil resource improvement and conservation. The success of these studies should be measured in terms of changes in the ability of soil to provide soil ES that contribute to human welfare.

REFERENCES

Bouma, J. 2001. The new role of soil science in a network society. Soil Science J. 166:12.

Bouma, J., Montanarella, L. 2016. Facing policy challenges with inter- and transdisciplinary soil research focused on the UN sustainable development goals. Soil J. 2:135145.

Boyd, J., Ringold, P., Krupnick, A., Johnston, R.J., Weber, M.A., Hall, K. 2014. Ecosystem services indicators: improving the linkage between biophysical and economic analyses. International review of environmental and resource economics. 8: 359–443.

Cohen, M.J., Brown, M.T., Shepherd, K.D. 2006. Estimating the environmental costs of soil erosion at multiple scales in Kenya using emerging synthesis. Agric. Ecosystem and Environment. 114,249–269.

Costanza, R., d'Arge, R., de Groot, R., Farberk, S., Grasso, M., Hannon, B., Limburg, K., Naeem, S., O'Neill, R.V., Paruelo, J., Raskin, R.G., Suttonkk, P., van den Belt, M. 1997. The value of the world's ecosystem services and natural capital. Nature, 387.

Daly, H.E., Farley, J. 2011. Ecological economics: principles and applications, second edition. Island Press, Inc.

de Groot, R., Brander, L., van der Ploeg, S., Costanza, R., Bernard, F., Braat, L., Christie, M., Crossman, N., Ghermandi, A., Hein, L., Hussain, S., Kumar, P., McVittie, A., Portela, R. Rodriguez, L.C., ten Brin, P., van Beukering, P. 2012. Global estimates of the value of ecosystems and their services in monetary units. Ecosystem Services, 1, 50–61.

de Groot, R.S., Wilson, M.A., and Boumans, R.M.J. (2002). A typology for the classification, description and valuation of ecosystem functions, goods and services. Ecological Economics, 41, 93–408.

Dominati, E., Patterson, M., Mackay, A. 2010a. A framework for classifying and quantifying the natural capital and ecosystem services of soils. Ecological Economics J. 69: 1858–1868.

Dominati, E., Patterson, M., Mackay, A. 2014b. A soil change-based methodology for the quantification and valuation of ecosystem services from agro-ecosystems: A case study of pastoral agriculture in New Zealand. Ecological Economics J. 100: 119–129.

Duffy, M. 2012. Value of soil erosion to the land owner. Ag Decision Maker, Iowa State University Extension Publication, A1–75.

Haab, T.C., Interis, M.G., Petrolia D.R., Whitehead J.C. 2013. From hopeless to curious? Thoughts on Hausman's "dubious to hopeless" critique of contingent valuation. Appalachian State University Department of Economics Working Paper. 13:07.

Hausman, J. 2012. Contingent valuation: from dubious to hopeless. Journal of Economic Perspectives 26:43.

Ingram, J., Mills, J., Dibari, C., Ferrise, R., Ghaley, B.B., Hansen, J.G., Iglesias, A., Karaczun, Z., McVittie, A., Merante, P., Molnar, A., Sánchez, B. 2016. Communicating soil carbon science to farmers: incorporating credibility, salience, and legitimacy. Journal of Rural Studies. 48:115–128.

Johnston, R.J., M. Russell. 2011. An operational structure for clarity in ecosystem service values. Ecological Economics J. 70:2243–2249.

Kling, C.L., Phaneuf, D.J., and Zhao J. 2012. From Exxon to BP: Has some number become better than no number? Journal of Economic Perspectives. 26:4.3–26.

Knight, Robins, and Chan, 2013. Natural capital: identifying implications for economics. HSBC Global Research.

Koch, A., McBratney, A.B., Adams, M., Field, D.J., Hill, R., Lal, R., Abbott, L., Angers, D., Baldock, J., Barbier, E., Bird, M., Bouma, J., Chenu, C., Crawford, J., Flora, C.B., Goulding, K.,Grunwald, S., Jastrow, J., Lehmann, J., Lorenz, K., Minansy, B., Morgan, C., O'Donnell, A.,Parton, W., Rice, C.W., Wall, D.H., Whitehead, D., Young, I., Zimmermann, M. 2014. Soil security: solving the global soil crisis. Glob. Policy J. 4:4.

Konarska, K.M., Sutton, P.C., Castellon, M. 2002. Evaluating scale dependence of ecosystem service valuation: a comparison of NOAA-AVHRR and Lands at TM datasets. Ecol.Econ. 41,491–507.

MEA. 2005. Millennium ecosystem assESment: ecosystems and human well-being: synthesis. Island Pres, Washington DC.

McBratney, A.B., Field, D.J., Koch, A. 2013. The dimensions of soil security. Geoderma. 213:203–213.

McBratney, A.B. 2016. Accepted, pending publication. The value of soil's contributions to ecosystem services. Global soil security report. Chapter 20.

Reimer, A., Thompson, A., Prokopy, L.S., Arbuckle, J.G., Genskow, K., Jackson-Smith, D., Lynne, G., McCann, L., Morton, L.W., Nowak, P. 2014. People, place, behavior, and context: A research agenda for expanding our understating of what motivates farmer's conservation behaviors. J. of Soil and Water Conservation. 69: 2.

Robinson, D.A., Lebron, L., Vereecken, H. 2009. On the definition of the natural capital of soils: a framework for description, evaluation and monitoring. Soil Sci. Soc. Am. J. 73:1904–1911.

Ruckelshaus, M., McKenzie, E., Tallis H., Guerry, A., Daily G., Kareiva P., Polasky, S., Ricketts, T., Bhagabati N., Wood S.A., Bernhardt, J. 2015. Notes from the field: lessons learned from using ecosystem service approaches to inform real-world decisions. Ecol. Econ. J. 115.11:21.

Smith, C.M, Williams, J.R., Nejadhashemi, A., Woznicki, S.A., Leatherman, J.C. (2014) Cost-effective targeting for reducing soil erosion in a large agricultural watershed. Journal of Agricultural and Applied Economics. 46:509–526.

Toman, M. 1998. Why not to calculate the value of the world's ecosystem services and natural capital. Ecol. Econ. J. 25:57–60.

Townsend, T.J., Ramsden, S.J., Wilson, P (2016). DAalyzing reduced tillage practices within a bio-economic modelling framework. Agricultural Systems 146. 91–102.

Soil security and policy

Global soil security for future generations

Luca Montanarella & Panos Panagos
European Commission

1 INTRODUCTION

Soil resources of our planet are limited and non-renewable in the timeframe of human generations. Therefore we need to secure sufficient soil resources for future generations if we want to achieve sustainability. Global governance of soil resources is necessary if we want to equitably share this precious limited resource. Attempts towards binding legal frameworks for protecting soils at regional or global scales have substantially failed, due to the strong National interests in maintaining full sovereignty on this fundamental resource for Nations. Soils are the foundation of Nations and have a strong link with the cultural and historical roots of each Nation in the world. Soil security can easily be perceived by National governments as a National security issue. On the contrary, the shared governance of the global soil resources calls upon sharing National soil resources with others, if we want to live on this planet in peace and security also in the future (Montanarella, 2015).

Sharing the governance of the National soil resources between Nations in a binding multilateral agreement is beyond reach in the current global political situation. What can be achieved is a voluntary process towards the common sustainable management of available soil resources. And this is indeed the way forward that has been chosen by Nations within the Global Soil Partnership (GSP) in 2011. The GSP is a voluntary partnership open to governments as well as to other stakeholders, like Non-Governmental Organizations (NGO's), scientific bodies and research institutions. It has by now successfully operated over more than 5 years and has already delivered major contributions towards sustainable soil management (SSM). The voluntary guidelines for Sustainable Soil Management (VGSSM) released in 2016 form the legal basis for governments to act within their territories towards soil security for sustainable development. The VGSSM are not binding but are formally endorsed by the FAO Council and therefore considered as agreed guidelines by all countries of the world, given the nearly universal membership in FAO (FAO, 2017). The scientific foundations for the VGSSM has been developed by the Intergovernmental Technical Panel on Soils (ITPS) established as the scientific body of the GSP. The ITPS has first completed in 2015 a full assessment of the global soil resources in the report "Status of World's Soil Resources" (FAO & ITPS, 2015). This report has clearly highlighted the urgent need to reverse the on-going soil degradation in all parts of the world. Major degradation processes identified by the experts are soil erosion, loss of soil organic carbon and nutrient imbalance in many soils of the world. Regional differences were highlighted in detailed regional assessments as annexes to the main report. On the basis of these findings, clear guidelines were drafted for approval by the FAO governing bodies resulting in the final version of the VGSSM. Full implementation of the VGSSM will ensure the achievement of global soil security, but will need the development of detailed implementation plans at regional and National scales, taking into account the various local soil conditions and differences in economic and social development. The GSP has developed by now a full set of regional implementation plans for each of the 5 pillars of action of the GSP:

2 SUSTAINABLE SOIL MANAGEMENT

This pillar is the core of the GSP, implying the actual implementation of the VGSSM on the ground. Full implementation of pillar 1 will imply achieving sustainable soil management and fully achieving soil security. This requires major investments in capacity building and technical support to land users for converting towards sustainable soil management practices.

3 AWARENESS RAISING AND EDUCATION

Only a small community of soil experts is actually fully aware of the challenges and threats of a declining soil resource at global level. Raising awareness and teaching to young and adults the fundamentals of soil functions and sustainable soil management is a pre-requisite for any further action towards global soil security. Building up a grassroots movement in all regions of the world advocating soil protection and SSM is necessary also for any further development of binding National or global soil protection policies and legislation. Without the support of the majority of the population towards SSM there will be little scope for further legislative action on soil protection.

4 RESEARCH AND DEVELOPMENT

Solid scientific evidence is needed in order to develop credible policy options for soil protection. The week scientific basis for many of the proposed measures in the past has been one of the causes of failure of some proposed legislation in the European Union for soil protection. The lack of transdisciplinary research on soil related issues, especially involving a full economic and social evaluation of the impact of proposed measures has been a major difficulty in getting soil protection measures accepted and implemented. Traditional soil science has been mainly focussing on physical-chemical properties of soils while neglecting soil biodiversity and the human dimension of soil management (economic and social aspects of soil degradation and restoration). The proposed strategy by the GSP for soil research, developing a truly innovative trans-disciplinary soil research agenda, will be a the core of the future activities of the GSP at global and regional level. Existing scientific communities and organizations can play a crucial role in this process. Especially the International Union of Soil Sciences (IUSS) has the potential for taking the lead in this process.

5 DATA AND INFORMATION SYSTEMS

Data and information on soils from the global to the local scales is a mandatory pre-requisite for any further assessment of soil resources and related policy interventions. Providing a solid evidence of data and information underpinning any proposal for sustainable soil management is necessary in order to avoid wrong and contradictory decisions. The current lack of reliable, updated, soil data and information in many parts of the world is one of the basic limitations for further action. Therefore the GSP has from its early beginning focussed a large part of its efforts in developing an updated and well-functioning global soil information system. Data on soils need to be consistent and comparable across scales as well as across continents and Nations. Important is as well the development of effective soil monitoring systems providing relevant data over long time frames in order to detect changes in soil condition of time. Soils are known for being highly resilient and most of their properties change rather slowly over long time frames. Assuring a long-term financial and technical support to soil monitoring systems has therefore always been a major difficulty. Examples of recently established soil monitoring systems are available in Europe (LUCAS-Soil) and USA (NRI).

6 STANDARDIZATION AND HARMONISATION OF METHODS

Data and information need to be comparable across various regions and Nations. Large differences in adopted standards and methods still exist across the world and are a major difficulty in acting together at global and regional level towards sustainable soil management. Developing common standards and methods within the GSP is therefore an important priority task. Most of the existing methodological differences are related to historical legacies of the past that should by now be overcome. The fact that the soil science community is still struggling in getting a single commonly accepted soil classification system is one of the examples of the remaining tasks for achieving a truly global approach to soil management. Attempts towards developing a common Universal Soil Classification System have so far failed. The GSP could certainly provide a valid framework in this rather difficult area of activity.

These 5 main pillars of action cover all 5 dimensions of global soil security as described by McBratney et al. (2014) (Fig. 1).

Figure 1. The 5 pillars of action of the GSP fully support the achievement of global soil security with a key role for pillar 1, Sustainable Soil Management and the related Voluntary Guidelines, fully including the 5 dimensions of soil security: Condition, Capability, Capital, Codification and Connectivity.

They as well address the goals and targets identified by the United Nations for achieving sustainable development. Achieving food security for all (SDG2), as well as a healthy live (SDG3) while preserving the terrestrial environment (SDG15) will require the full implementation of the VGSSM. Restoring large areas of degraded land is a mandatory requirement for achieving a land-degradation neutral world, as advocated by Target 15.3 of SDG 15. Adopting sustainable soil management practices in large areas currently used for intensive agricultural production will decrease the environmental and human health impact of food production, securing enough healthy food for all of us. Those practices are clearly spelled out in the VGSSM and will need to be supported by the necessary local extension services providing the technical support to farmers for implementing such innovative techniques. The transition from an unsustainable business-as-usual scenario to a more sustainable land use scenario will imply a cost that needs to be spread on all citizens given the collective interest of all of us in maintaining a sustainable and healthy soil management system. Land owners need to be supported by public funding for completing the transition to a more sustainable soil management system. In Europe the Common Agricultural Policy (CAP) provides already major incentives towards good agricultural practices that increase soil protection from erosion and organic carbon depletion (Panagos et al., 2015). There is the need to move on towards a full implementation of the VGSSM also in Europe as a major component of the revised CAP.

In conclusion, achieving global soil security for future generations is possible if we start already now in fully implementing the policies and scientific knowledge we have for sustainable soil management. The GSP provides the necessary enabling framework for the full implementation of the VGSSM and will be instrumental in achieving the soil related SDGs by 2030.

REFERENCES

FAO and ITPS, 2015. Status of the World's Soil Resources (SWSR) – Main Report. Food and Agriculture Organization of the United Nations and Intergovernmental Technical Panel on Soils, Rome, Italy.

FAO 2017. Voluntary Guidelines for Sustainable Soil Management, Food and Agriculture Organization of the United Nations, Rome, Italy.

McBratney, A., Field, D.J. and Koch, A., 2014. The dimensions of soil security. Geoderma, 213, 2013–213.

Montanarella, L. 2015. Agricultural policy: Govern our soils. Nature, 528 (7580), pp. 32–33.

Panagos, P., Borrelli, P., Robinson, D.A. 2015. Common Agricultural Policy: Tackling soil loss across Europe. Nature, 526 (7572), p. 195.

Contribution of knowledge advances in soil science to meet the needs of French state and society

V. Antoni, H. Soubelet & G. Rayé
French ministry of Ecological and Solidarity Transition, France

T. Eglin, A. Bispo & I. Feix
French Environment and Energy Management Agency, France

M.-F. Slak
French Ministry of Agriculture and Food, France

J. Thorette
Centre-Val de Loire directorate of Environment, Land Planning and Housing, France

J.-L. Fort
Nouvelle Aquitaine Chambre of Agriculture, France

J. Sauter
Association for Agricultural Revival in Alsace, France

ABSTRACT: In the latest 1990's, several initiatives launched continued partnerships between French administration and soil science sphere. The Group of scientific interest on soils, Gis Sol, created in 2001, was an ambitious challenge to improve soil knowledge and understanding. This group has published the first appraisal of soil quality in France.

In parallel, the French ministry of Environment has funded the Gessol research programme. It succeeded in structuring a multi-disciplinary soil research community in France, providing policy makers and land users knowledge and operational tools to manage natural, agricultural or urban soil quality.

Both programmes initiated proximity between public funders and soil scientists.

1 INTRODUCTION

In the latest 1990's, several initiatives launched efficient and continued partnerships between the French administration and the soil science sphere. Indeed, except for the fields of contaminated sites and sewage sludge spreading, there was a great lack in soil knowledge and understanding. Moreover, the scientific community was not structured in order to meet the needs of French policy makers regarding soil-related issues.

2 FRENCH SOIL KNOWLEDGE PILLARS

2.1 *First pillar: The French soil inventory and monitoring programmes*

The **French Soils Scientific Interest Group (Gis Sol)**, see website www.gissol.fr) was created in 2001. It involved catching up, over a few years, with France's lateness in carrying out inventory and monitoring programmes of its soils, and making soil data available to policy-makers and the general public. Thus, ministries in charge of Agriculture and Environment, the main research organizations and public agencies involved in soil science gathered around a common purpose: to equip France with a national information system regarding soils, focused on their inventory and monitoring (properties and quality). After fifteen years, the Gis Sol notably improved the knowledge on French soil. For example, the French soil quality monitoring network, also known as RMQS (Réseau de mesures de la qualité des sols), has allowed to spatially assess soil quality across mainland France and overseas territories. Figure 1 shows for instance the huge improvements of knowledge on soil mercury concentrations in mainland France between 2001 and 2017. Different productions have besides permitted to disseminate this knowledge to a broad

(a)

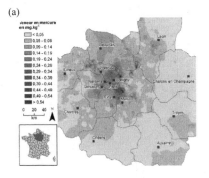

(b)

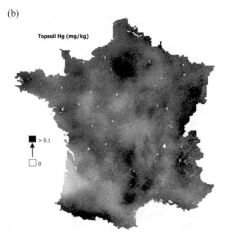

Figure 1. From a mercury topsoil amount mapping around Paris in 2001(Source: Ademe-INRA, BD-ETM, 2001; IGN, Geofla®, 2008) – (a) to a robust geostatistical prediction of mercury concentrations across France in 2017 (Marchant et al. 2017, sources: Gis Sol/RMQS) (b).

audience: publication of the first report on the state of the soils in France, whose synthesis was translated to English (Gis Sol 2011; Gis Sol 2013); enhancement of the Gis Sol website (see www.gissol.fr), proposal of latest technology to access soil data (at least fifty webservices available), online mapping interface; organization of regular soil conferences for a wide range of end-users; data supply to the French ministries for their own publications at national (CGDD/SOeS 2015; CGDD/SOeS 2014) or at regional levels, like several soil leaflets published in the framework of the "French regional environmental profiles" (Dreal Centre-Val de Loire 2015; Gip Bretagne Environnement 2016; Dreal Grand Est, Dreal Hauts-de-France, Dreal Occitanie, Dreal Pays de la Loire: online soil webpages); management and storage of soil samples in the European soil sample archive. The Gis Sol also provides some data or indicators needed to report on French soil quality at a European level in the scope of the European environment information and observation network (Eionet) or the European Commission.

2.2 *Second pillar: The French applied research programmes*

In 1998, the French ministry of Environment has set up a research programme entitled **Environmental functions and management of soil heritage (Gessol**, see website www.gessol.fr). This multidisciplinary research programme aimed to work on all the soil functions by the mobilization of a large panel of sciences disciplines including social sciences. It focused on structuring a soil research community providing policy makers and land users with knowledge and operational tools to assess, monitor, manage or improve soil quality, either in natural, agricultural or urban contexts. Between 1998 and 2014, Gessol launched three research calls for tender. The research priorities were selected by end-users, decision makers and environmental managers concerned with soils (Breure et al. 2012). In the end, the Gessol research programme has financed 46 projects on soils. In particular, these projects have contributed to the implementation and improvement of soil observation systems, have produced recommendations of environment-friendly soil management practices and have developed environmental assessment methodologies (Citeau et al. 2010; Bispo et al. 2016), in a multifunctional perspective. Gessol has also contributed to arise awareness around the necessity to preserve soils by supporting the creation and dissemination of playful tools (The Hidden life of soils – http://www.gessol.fr/game-hidden-life-soils) and pedagogical booklets (CGDD 2010) focusing on the hidden life of soils.

More recently, new initiatives in soil science occurred. For instance, the **Bioindicators – biological tools for sustainable soils – programme** (Peres et al. 2011; Peres et al. 2012; Bispo et al. 2009) funded by the French environment and energy management agency (Ademe – Agence de l'environnement et de la maîtrise de l'énergie) promoted the development and the standardization of bioindicators to monitor soil quality, to characterize thoroughly biological state of French soils and, finally, to assess the risks for ecosystems and polluted sites (Ademe, 2011). In 2010, Ademe also launched a research funding programme called **Reacctif (research for agricultural and forestry mitigation of climate change)** to provide a better understanding of soil and climate interactions, mainly through greenhouses gases (GHSs) fluxes. The Reacctif programme has funded 24 projects focusing on soil-climate interactions. In particular, these projects contribute to assess the mitigation potential of agricultural and forest soils management practices (e.g. Epron et al. 2016; Peyrard et al. 2016; Cardinael et al. 2015), to a better representation of the territorial diversity of soil GHGs emissions and carbon stocks, and to the development of methods for diagnosing GHGs emissions balances, integrating soil carbon balances (e.g. Sierra et al. 2017).

The French ministry of Agriculture also supports applied research on soils through the special allocation account for agricultural and rural development (CASDAR), used for funding i) applied research projects, with almost 50 soil-related projects funded over the 2004–2016 period, accounting for more than 20 million € (Maaf 2016); ii) the coordination of the **Combined technology network on soils and territories (RMT Sols et territoires**, see http://www.sols-et-territoires.org/) and by launching the "4 per 1000" initiative (see http://4p1000.org/understand): soils for food security and climate (Inra, IRD, Cirad, Maaf 2015). The first project aims to increase and enhance the knowledge on soils and their use by a wide range of decision-makers at the territorial level. To achieve this, it fosters interactions between organizations involved in agricultural and rural development, agricultural education institutions and research institutes, and acts as an incubator for new projects or as an amplifier in the dissemination of decision-support tools. The second project aims to improve the organic matter content and to promote carbon sequestration in soils through the application of adapted agricultural practices. This "4 per 1000" initiative results from the advances of the soil science research on agroecology, agroforestry and conservation agriculture.

3 THE FIVE DIMENSIONS OF FRENCH SOILS' SECURITY

A general review of knowledge advances in soil science in France since the latest 90's has led to a tremendous account of scientific publications, at least 30 seminars organized around soils, 40 information letters (La lettre du Gis Sol), a national information system dedicated to soils (Gis Sol/Donesol), 50 webservices (Gis Sol), 3 websites totally dedicated to the main soil programmes (Gis Sol, Gessol, RMT Sols et Territoires) and pedagogical material (Gessol, Gis Sol, Ademe).

The relationships with the five dimensions of soil security depend on the nature of the French soil programmes: Gis Sol (capability, condition, connectivity, codification), Gessol (capability, capital, connectivity), RMT Sols et Territoires (capability, connectivity), Ademe programmes (capability, capital, connectivity).

3.1 Assessment of French soils' capability

France made very strong advances in the knowledge of soil geographical distribution during the last decades. This is mainly due to soil mapping and monitoring activities undergone under the Gis Sol framework.

In order to answer the question "What can this soil do?", the first two pillars, namely Gis Sol and Gessol programmes, were launched in the latest 1990's. Building an efficient French Soil information system on the one hand, and performing a multidisciplinary research programme based on a large panel of scientific disciplines on the other hand, were the main aims of those programmes.

At the same period, the implementation of the Kyoto protocol has put an emphasis on the capability of soils to contribute to climate change mitigation. On the request of the French ministry of Environment, a **collective scientific expertise on soil carbon** (Arrouays et al. 2002) provided a first assessment report on the capacity for organic carbon accumulation in French soils. With the "4 per 1000" initiative and the aim of France to reach carbon neutrality by 2050 (see French climate plan: https://www.ecologiquesolidaire.gouv.fr/planclimat), there is a renewed interest for a better assessment of the potential of organic carbon sequestration in soils. In 2017, Ademe and the French ministry of Agriculture requested French institute for agronomical research (Inra) an update of the Arrouays et al. assessment. The study is expected to be completed by the end of 2018 and should provide carbon sequestration potential related to soil management in France, and a methodological framework that could be adapted in other countries.

However, there is still a lack of information at detailed scale, where action is necessary. More detailed information about soil capability would necessitate boosting soil mapping activities, mainly through the combination of high resolution digital soil mapping with further field surveys.

3.2 Assessment of French soils' condition

A first assessment of soils' condition has been done thanks to soil monitoring programmes i.e. RMQS and the soil-test analysis databases (BDAT – Base de données des analyses de terre—on agronomical parameter such as pH or carbon; BDETM – Base de données des éléments traces métalliques—on trace elements). The baseline of some dynamic soil properties is now well characterized. Some trends in soil properties have also been assessed. The second round of the RMQS will give information on trends in many components of soils' condition.

French soils' condition is monitored by a set of indicators in the Gis Sol framework. Amongst them, the most emblematic refer to carbon stocks, pH, soil texture, available water capability, soil erosion risk, soil biodiversity index.

As previously described, in 2011, the Gis Sol has published the first report on the state of the soils in France (Gis Sol 2011; Gis Sol 2013, see https://www.gissol.fr/rapports/synthesis_HD.pdf). It describes all those indicators, highlighting the French soil diversity and its evolution under different pressures. This assessment, based on the French soil quality monitoring programmes, is intended to become a reference tool in support of the emergence of a sustainable soil management strategy. In addition, on the occasion of the International Year of Soils in 2015, the ministry in charge of Environment published a booklet of about 50 indicators on soils depicting the principal key figures (CGDD/SOeS 2015).

The investigations in the field of soil carbon stocks (Martin et al. 2011) have fed the GHGs French inventory (Citepa, 2017), while the one referring to soil biodiversity such as DNA amounts (Ranjard et al. 2010; Ranjard et al. 2012) abundance and number of earthworms taxons in French soils (Peres et al. 2012) were added to the **French biodiversity observatory (ONB**, see http://indicateurs-biodiversite.naturefrance.fr/en), which aims to provide robust indicators in the scope of the French strategy for biodiversity (MEDTL 2010). Dedicated to agriculture, the **Agricultural biodiversity observatory (OAB**, see website: http://observatoire-agricole-biodiversite.fr/) has two objectives: i) to fill in a scientific database in order to compute indicators on ordinary biodiversity in agricultural lands and to analyze the link between biodiversity and agricultural practices; ii) to arise awareness and to support stakeholders (farmers, etc.) about the role of agricultural practices. This observatory was created at the initiative of the French ministry of Agriculture and coordinated from a scientific point of view by the French Museum of Natural History. Results are published every year (e.g. OAB 2014).

More recently, in 2016, the French ministry of Environment cofinanced and copiloted with the French ministry of Agriculture and Ademe, a **collective scientific expertise on soil sealing** (Inra et Ifsttar, in prep.), entrusted to the Inra and the French institute of science and technology for transport, spatial planning and networks (Ifsttar). This expertise focuses on several issues regarding soil sealing: state of soils, soil sealing trends, determinants of soil sealing, environmental, social and economic impacts, levers, and research priorities.

At a European level, the indicators reported by France towards different instances (OCDE, Eurostat, Eionet, CE-DGEnv) are aggregated at a national level or by NUTS1 essentially from the French indicators developed in the frame of the Gis Sol. For instance, among the Eurostat-OCDE agro-environmental indicators, the figures reported for lands affected by water erosion in 2000, 2006 and 2012 concerned on the one hand different land categories (total agricultural land, arable and permanent crop land, permanent meadows and pasture), and on the other hand several erosion rates.

3.3 *Forward to assess French soils' capital*

Some elements referring to capital are already available. A good example is the assessment of carbon stocks in soils. Some research about the economic value of soils have also been conducted in the framework of the Gessol research programme.

Indeed, a soil natural capital and ecosystem services approach enables the contribution of soils to human well being to be recognised in societal decision making (Breure et al. 2012). Recently, the **French assessment of ecosystems and ecosystem services (Efese**, see www.ecologique-solidaire.gouv.fr/levaluation-francaise-des-ecosystemes-et-des-services-ecosystemiques) managed by the French ministry of Environment included some soil-related ecosystem services indicators based for instance on erosion and soil carbon modelling (Inra, in prep.).

3.4 *French soils' connectivity*

All three Gis Sol, Gessol and Ademe programmes initiated proximity between public funders and soil scientists through exchange forums such as scientific councils, seminars, etc.

Many activities allowed to improve the connectivity between soil and local actors. The Gis Sol for

(a) (b)

Figure 2. (a) Gis Sol 2013. The state of the soils in France. A synthesis. Groupement d'intérêt scientifique sur les sols, France, 24 p. (b) CGDD/SOeS 2015. Sols et environnement: chiffresclés, édition 2015. Collection Repères. 104 p.

instance published a methodological guide (Medde et Gis Sol 2013) in order to help field experts to define wetlands as required by the law for the purpose of safeguard (French ministerial Decree of 24 June 2008 amended). Indeed, the identification criteria of those sensitive environments include soil criteria in addition to vegetation ones.

In this instance, the RMT Sols et Territoires (see website http://www.sols-et-territoires.org/) emerged in 2010 around Inra (Infosol unit) and a core group of organizations with a regional pedological expertise, with two main objectives: to improve soil knowledge at the territorial level and to promote the use of soil data in various soilrelated decision-making. This group developed several actions intended to land planners, agricultural or educational professionals in order to arise awareness about soils and to disseminate some useful tools to lead to local development projects. Among the potential beneficiaries of the works developed in the RMT Sols et Territoires framework, the farmers are obviously targeted.

For instance, the TYPTERRES guide (http://www.sols-et-territoires.org/produits-du-reseau/projets-affilies-au-rmt-st/typterres/), product of a dialog between soil scientists and agronomists, offers a method to build an agronomical typology of soils. It is based on soil inventories at the 1 to 250,000 scale established in the frame of the Gis Sol (Inventory management and soil conservation programme—IGCS). Thanks to these typologies which provide soil data for indicators, models or decision-making tools can be shared by all the stakeholders involved in agronomical consulting or agroenvironmental assessment in several topics, such as: risks of nitrogen leakage into water or air, or even pesticides, dose of nitrogen, water balances and irrigation management. Another example is the ABC'Terre project that has developed a method allowing assessment of the soil organic carbon status according to various agricultural scenarios at the territorial level (see an application regarding soil erosion management in Van DijK et al. 2016). All those initiatives and tools focus on helping soil managers in taking soil into account in their activities.

Various pedagogical documents have also been published recently by the RMT Sols et Territoires like a text-book on soil data (Ducommun et al. 2017). It highlights available spatial soil data and their usefulness in various contexts, through the development of case studies and exercises for teachers and students.

Some communication actions clearly targeted the citizens or the policy makers. Pedagogical materials have indeed been published in order to disseminate knowledge on soils towards the society. For instance, the Happy Families card game

Figure 3. Book "Maps and pedological data: tools for the territories" published in 2017 (RMT Sols et Territoires, Collectif, Ducommun et al. 2017).

Figure 4. Exemple of the mesofauna family in the midst of the Happy Families card game "The hidden life of soils" (Gessol 2010).

"The hidden life of soils" (Gessol 2010), allows to have fun learning about the diversity and functioning of the living soil environment from the 42 playing cards, each with a large picture and description. A supplemental educational booklet is also included for a deeper understanding. Those pedagogical tools were translated in English and Portuguese to ensure a wider spreading and can freely be downloaded (www.gessol.fr).

Eventually, big progress has been made to connect the policy makers and the stakeholders to soil.

3.5 *A way towards better codification?*

A logical consequence of the better connectivity of policy makers to soil should be the increase of soil related policies in France.

So far, both ministries in charge of Environment or Agriculture rely on soil science knowledge to ensure decision-making tools to environmental (including urban planners) or agricultural stakeholders. For instance in 2008, the French ministry of Agriculture asked Inra to assess projections of GHGs emissions and removals from the land use, their changes and forest sector (LULUCF) by 2020 in France (Forslund et al. 2009).

Besides, in 2010, the French ministry of Environment asked the Gis Sol to launch a decision-tool for the contaminated soils and sites' stakeholders. Based on all nine heavy metals amounts measured in the RMQS programme (Cd, Co, Cr, Cu, Mo, Ni, Pb, Tl, Zn), indicators assessing soil pedogeochemical background were thus provided on the Gis Sol website (Saby et al. 2009).

At last, the French legislation imposes to delineate erosion risk zones (Law n°2003–699, Application Decree n°2005–117). A methodological guide for a departmental zoning of soil water erosion risk was thus published at the request of the French ministry of Environment. It shows how to process from soil inventories in two local case studies (Desprats et al. 2006, Surdyk et al. 2006) and provides general recommendations as well (Cerdan et al. 2006).

Applied research is also conducted to better evaluate and manage soil potentialities within land-use planning policies. For example, the main ambition of one Gessol project, namely Uqualisol-ZU (Keller et al. 2012), was to put urban planning laws into perspective with the scientific knowledge of soil functions. Keller et al. proposed an index displaying how relevant land use is, given the multiple potentialities of the soil. This index was tested in the periurban context of the mining basin of Provence, near the regional metropolis of Aix-Marseille-Provence in the south of France.

At a pan-European level, Gessol has also contributed to the discussions preparing the European Strategy for soil protection. Moreover, the European Commission (regulation UE n°1305/2013 on support for rural development by the European Agricultural Fund for Rural Development—Feader) wished to revise and to harmonize the delimitation of agricultural areas subject to natural constraints (Jones et al. 2014). Each Member State is therefore bound to produce this new delineation on the basis of biophysical and socio-economic criteria defined by the Commission. Soil data at 1 to 250,000 scale, established under the framework of the Gis Sol (IGCS programme), are therefore used to assess biophysical criteria. This work is still in progress, managed by the French ministry of Agriculture, with technical support from Inra.

Finally, France is involved in the ISO Technical Committee (TC) 190 "Soil quality". International standardization is a way to share knowledge and to reach agreement on methodological documents (e.g. standard, guideline, technical report) that can be used worldwide. Standards can also provide support to implement public policies. Recently, the main emphasis concerned the analysis of soil contaminants and their impact on soil living organisms. ISO TC 190 also develops document to implement monitoring programmes and to measure soil carbon stocks and GHGs emissions (Bispo et al. 2017).

4 OUTLOOKS

4.1 *French strategy for good soil management*

Both ministries of Agriculture and Environment jointly explore the possibility to elaborate a French strategy for good soil management with the goal of achieving of a greater consistency of public action for soil conservation in terms of quantity, of quality, of functions and of eco-system services. The improvement of knowledge on soils in order to facilitate their sustainable management and the arising of education and awareness of stakeholders and society to the issues concerning soils are also taken into account in this strategy.

4.2 *French network of scientific and technical expertise on soils*

Under the Agriculture-Innovation 2025 plan, the **French network of scientific and technical expertise on soils (RNEST)** has been launched in December 2016. This networking should contribute to develop transdisciplinary applied research on soils, as well as to strengthen the international visibility of French expertise in this field. Moreover, the establishment of the RNEST should make it possible to increase soil related policies in France in the next coming years.

5 CONCLUSION

In the end, all those programmes linking the environmental and agricultural public organizations on the one hand, and the soil scientists on the other hand, allowed France to create without doubt an efficient network entirely focused to heighten awareness of all society' actors to the main significance of soils.

ACKNOWLEDGEMENTS

We warmly thank all the stakeholders involved since years in the enhancement of soil knowledge in France. Among them, the main funders (French ministries of Agriculture and Environment, the French environment and energy management Agency—Ademe), the research organisms (the French Institute for Agronomic Research—Inra, the French Institute for Research and Development—IRD, etc.), the regional partners (Chambers of agriculture, agricultural and technical institutes, agronomic high schools, etc.).

A special thank goes naturally to all the soil surveyors and technical assistants involved in sampling and inventorying the soils, without whom all those marvelous projects couldn't have seen the day.

We thank also to the French association for soil studies (Afes, see website http://www.afes.fr/), the learned society devoted to soil, member of the International soil science association, for allowing since 1934 soil scientists to be connected together.

REFERENCES

Ademe 2011. Bioindicateurs pour la caractérisation des sols. Journée technique nationale, Paris 7e. Recueil des interventions.

Arrouays D. et al. 2002. Contribution à la lutte contre l'effet de serre. Stocker du carbone dans les sols agricoles de France? Expertise scientifique collective. Rapport. Inra. 332 p.

Bispo, Antonio, Lizzi Andersen, Denis A. Angers, Martial Bernoux, Michel Brossard, Lauric Cécillon, Rob N. J. Comans et al. 2017. Accounting for Carbon Stocks in Soils and Measuring GHGs Emission Fluxes from Soils. Frontiers in Environmental Science 5, doi:10.3389/fenvs.2017.00041.

Bispo A., Guellier C., Martin É., Sapijanskas J., Soubelet H. et Chenu C., coord. 2016. Les sols: intégrer leur multifonctionnalité pour une gestion durable. Collection Savoir Faire, Editions Quae, 384 p.

Bispo A., Grand C. et Galsomies L. 2009. Le programme Adema "Bioindicateurs de qualité des sols": vers le développement et la validation d'indicateurs biologiques pour la protection des sols. Étude et Gestion des Sols, Volume 16, 3/4, 2009 - pages 145 à 158.

Breure A.M., De Deyn G.B., Dominati E., Eglin T., Hedlund K., Van Orshoven J. et Posthuma L. 2012. Ecosystem services: a useful concept for soil policy making. Current opinion in Environmental Sustainability, 4, 578–585.

Cardinael, R., Chevallier, T., Barthes, B., Saby, N., Parent, T., Dupraz, C., et al. 2015. Impact of alley cropping agroforestery on stocks, forms and spatial distribution of soil organic carbon – a case study in a Mediterranean context. Geoderma 259, 288–299. doi: 10.1111/gcb.13244.

Cerdan O, Le Bissonnais Y, Souchère V, King C, Antoni V, Surdyk N, Dubus I, Arrouays D et Desprats JF 2006. Guide méthodologique pour un zonage départemental de l'érosion des sols. Rapport n°3: synthèse et recommandations générales. BRGM/RP-55104-FR. 87 p.

CGDD 2010. La vie cachée des sols: l'élément essentiel d'une gestion durable et écologique des milieux. 20 p.

CGDD/SOeS 2015. Sols et environnement: chiffresclés, édition 2015. Collection Repères. 104 p.

CGDD/SOeS 2014. L'environnement en France—Édition 2014. Chapitre « Les sols ». Collection Références.104 p., pp. 77–88.

Citeau L., Bispo A., Bardy M. et King D., coord. 2008. Gestion durable des sols. Collection Savoir Faire, Editions Quae, 320p.

Citepa 2017. Rapport Ominea—Organisation et méthodes des inventaires nationaux des émissions atmosphériques en France. (https://www.citepa.org/images/III-1_Rapports_Inventaires/OMINEA_2017.pdf). 838p.

Desprats J.F., Bourguignon A., Cerdan O., Le Bissonnais Y., Colmar A. 2006. Guide méthodologique pour un zonage départemental de l'érosion des sols. Rapport n°1: Etude de sensibilité sur le département de l'Hérault Rapport BRGM-RP-55049, 67 p.

Dreal Centre-Val de Loire 2015. Le sol, des avancées dans la connaissance. Collection Les synthèses du profil environnemental régional. 13 p.

Dreal Centre-Val de Loire 2015. Sol et soussol: supports des activités humaines. Collection Les cahiers cartographiques du profil environnemental régional. 14 p.

Dreal Grand Est: http://www.per.alsace.developpementdurable.gouv.fr/accueil/thematiques_environnementales/sols_et_sous_sols.

Dreal Hauts-de-France: http://www.hauts-de-france.developpement-durable.gouv.fr/?-Sols-et-sous-sols-.

Dreal Occitanie: http://www.occitanie.developpement-durable.gouv.fr/sols-et-sous-sols-r1603.html.

Dreal Pays de la Loire: http://www.profil-environnemental.pays-de-la-loire.developpement-durable.gouv.fr/sol-et-sous-sol-r41.html.

Epron, D., Plain, C., Ndiaye, F. K., Bonnaud, P., Pasquier, C. et Ranger, J. 2016.Effects of compaction by heavy machine traffic on soil fluxes of methane and carbon dioxide in a temperate broadleaved forest. For. Ecol. Manage.382, 1–9. doi: 10.1016/j.foreco.2016.09.037.

Forslund A., Colin A., De Cara S., Leban J.-M., Martin M., Mathias E., Guyomard H. et Stengel P. 2009. Projections d'émissions et d'absorptions de gaz à effet de serre du secteur utilisation des terres, leurs changements et la forêt (UTCF) à l'horizon 2020 en France, rapport final pour le ministère de l'Agriculture, de l'Alimentation et de la Pêche, INRA-IFN-CITEPA, Paris, France, 142 p.

Gip Bretagne Environnement 2016. La biodiversité des sols bretons. Dossier n°12, novembre 2016. 13 p.

Inra et Ifsttar, in prep. 2017. Artificialisation des sols: déterminants, impacts et leviers d'action. Synthèse du rapport de l'Expertise scientifique collective. To be published in 2017.

Gessol: http://www.gessol.fr.

Gessol 2010. The Hidden life of soils: http://www.gessol.fr/content/le-jeu-de-7-familles-la-vie-cach-e-des-sols.

Gis sol: https://www.gissol.fr/.

Gis Sol 2013. The state of the soils in France. A synthesis. Groupement d'intérêt scientifique sur les sols, France, 24 p.

Gis Sol 2011. L'état des sols de France. Groupement d'intérêt scientifique sur les sols, 188 p.

Inra in prep. Etude Inra «EFESE-écosystèmes agricoles», rapport final.

Inra et Ifsttar in prep. Artificialisation des sols: déterminants, impacts et leviers d'action Synthèse du rapport de l'Expertise scientifique collective.

Jones R.J.A., Van Diepen K., Van Orshoven J., Confalonieri Roberto 2014. Scientific contribution on combining biophysical criteria underpinning the delineation of agricultural areas affected by specific constraints. Scientific and Technical Research Reports: Report EUR 26940 EN. EC-JRC Publications Office. 85 p.

Keller C., Ambrosi JP., Rabot E., Robert S., Lambert M.L., Criquet S., Ajmone Marsan F. et Biasioli M. 2012. Soil Quality Assessment for Spatial Planning in Urban and Peri-Urban Areas – Municipalities of Gardanne and Rousset (southern France). Bulletin of the European Land and Soil Alliance Local land et soil news, 40/41: 12–14.

Maaf 2016. La recherche appliquée sur les sols soutenue par le CASDAR, 2004–2016. 50 p.

Maaf 2015. Join the 4 per 1000 Initiative: soils for food security and climate. 8 p.

Marchant B.P, Saby N.P.A. et Arrouays D. 2017. A survey of topsoil arsenic and mercury concentrations accross France. Chemosphere 181 (2017) 635–644.

Martin M.P., Wattenbach M., Smith P., Meersmans J., Jolivet C., Boulonne L. et Arrouays D. 2011. Spatial distribution of soil organic carbon stocks in France. Biogeosciences, 8, 1053–1065.

MEDDE et Gis Sol. 2013. Guide pour l'identification et la délimitation des sols de zones humides. Ministère de l'Écologie, du Développement Durable et de l'Énergie, Groupement d'Intérêt Scientifique Sol, 63 pages.

MEDTL 2010. Stratégie nationale pour la biodiversité 2011–2020. Texte intégral. 60 p. (see http://www.naturefrance.fr/).

MEDTL 2010. Stratégie nationale pour la biodiversité 2011–2020. Les 20 objectifs de la SNB. 1 p. (see http://www.naturefrance.fr/).

Peyrard C., Mary B., Perrin P., Véricel G., Gréhan E., Justes E. et Léonard J. 2016. N2O emissions of low input cropping systems as affected by legume and cover crops use. Agriculture, Ecosystems and Environment. 224: 145–156. DOI: 10.1016/j.agee.2016.03.028.

OAB 2014. Observatoire agricole de la biodiversité. Lettre d'information spéciale n°18 – BILAN 2014. 11 p.

ONB: http://indicateurs-biodiversite.naturefrance.fr/en.

Pérès G., Vandenbulcke F., Guernion M., Hedde M., Beguiristain T., Douay F., Houot S., Piron D., Richard A., Bispo A., Grand C., Galsomies L., Cluzeau D. 2011. Earthworm indicators as tools for soil monitoring, characterization and risk assessment. An example from the national Bioindicator programme (France). Pédobiologia. Volume 54, p 77–87.

Peres G., Bispo A., Grand C. et Galsomies L. 2012. Le programme de recherche Ademe "Bioindicateurs de l'état biologique des sols". Ses objectifs, sa mise en œuvre et son déroulement. 26 p.

RMT Sols et Territoires: http://www.sols-et-territoires.org/.

Ranjard L., Dequiedt S., Chemidlin Pre´vost-Boure N., Thioulouse J., Saby N.P.A., Lelievre M., Maron P.A., Morin F.E.R., Bispo A., Jolivet C., Arrouays D. et Lemanceau P. 2013. Turnover of soil bacterial diversity driven by wide-scale environmental heterogeneity. Nature communication, 4:1434, DOI: 10.1038/ncomms2431.

Ranjard L., Dequiedt S., Jolivet C., Saby N.P.A., Thioulouse J., Harmand J., Loisel P., Rapaport A., Saliou Fall, Simonet P., Joffre R., Chemidlin-Pr´evost Bour´e N., Maron P.A., Mougel C., Martin M.P., Toutain B., Arrouays D., Lemanceau P., 2010. Biogeography of soil microbial communities: a review and a description of the ongoing French national initiative. Agron. Sustain. Dev. 30 (2010) 359–365. DOI: 10.1051/agro/2009033.

RMT Sols et Territoires 2017. Collectif, coordination Ducommun Ch. En collaboration avec Lucot E. Les cartes et les données pédologiques, des outils au service des territoires. Educagri éditions, Dijon. 172 p.

Saby N.P.A., Thioulouse J., Jolivet C.C., Ratié C., Boulonne L., Bispo A. et Arrouays D. 2009. Multivariate analysis of the spatial patterns of 8 trace elements using the French soil monitoring network data. Science of the Total Environment 407 (2009) 5644–5652.

Sierra J., Causeret F. et Chopin P. 2017. A framework coupling farm typology and biophysical modelling to assess the impact of vegetable crop-based systems on soil carbon stocks. Application in the Caribbean. Agricultural Systems 153:172–180 doi: 10.1016/j.agsy.2017.02.004.

Surdyk N., Cerdan O. et Dubus I.G. 2006. Guide méthodologique pour un zonage départemental de l'érosion des sols. Guide méthodologique pour un zonage départemental de l'érosion des sols. Rapport n°2: Etude de sensibilité sur le département de l'Oise. BRGM/RP-55103-FR, 63 p.

Van Dijk P., Rosenfelder C., Scheurer O., Duparque A., Martin P. et Sauter J. 2016. Une approche agronomique territoriale pour lutter contre le ruissellement et l'érosion des sols en Alsace. Agronomie, Environnement, Sociétés 6:1–5.

Soil security and mapping

GlobalSoilMap and the dimensions of Global Soil Security

Dominique Arrouays & Anne C. Richer-de-Forges
INRA, InfoSol Unit, Orléans, France

Alex B. McBratney & Budiman Minasny
University of Sydney, Sydney, Australia

Mike Grundy & Neil McKenzie
CSIRO, Australia

Zamir Libohova
USDA, NRCS, Lincoln, Nebraska, USA

Pierre Roudier
Landcare Research—Manaaki Whenua, Manawatū Mail Centre, New Zealand

Jon Hempel[†]
Retired, USDA, NRCS, Lincoln, Nebraska, USA

ABSTRACT: The demand for global information on functional soil properties is high and has increased over time. Responding to these challenging demands requires relevant, reliable and applicable information. The *GlobalSoilMap* initiative is a response to such a soaring demand for up-to-date and relevant soil information. Ultimately, the *GlobalSoilMap* grid will provide suitable data to a wide variety of communities that makes decisions at various levels from local (field) to national scale and beyond. *GlobalSoilMap* is contributing to soil security assessment mainly through its contribution to capacity and condition assessment. Moreover, it may bring some inputs to the other dimensions of soil security. In particular, its contribution to soil capital assessment may re-inforce soil awareness to policy makers, soil managers and citizens and therefore increase the soil connectivity and codification dimensions.

1 INTRODUCTION

Unprecedented demands are being placed on the world's soil resources (Hartemink & McBratney, 2008; Amundson et al. 2015; Montanarella et al. 2016). Indeed, the demand for global information on functional soil properties is high and has increased over time. Responding to these challenging demands requires relevant, reliable and applicable information (Koch et al. 2013). The *GlobalSoilMap* consortium was established in response to such a soaring demand for up-to-date and relevant soil information (Sanchez et al. 2009; Arrouays et al. 2014). The majority of the data needed to produce *GlobalSoilMap* soil property maps will, at least for the first generation, come mainly from archived soil legacy data, which could include polygon soil maps and point pedon data (Arrouays et al. 2017a), and from available co-variates such as climatic data, remote sensing information, geological data, and other forms of environmental information. The predictions and estimations are generated using state-of-the-art Digital Soil Mapping techniques (McBratney et al. 2003; Minasny & McBratney, 2016; Arrouays et al. 2017b).

Procedures and methodologies to produce this information vary depending on the types and amount of available data and expertise, but all information meet a specific set of standards by the *GlobalSoilMap* specifications (*GlobalSoilMap*, 2015). These specifications are the only specifications for a global soil information system to be derived using a consensus-based process. They have been recently endorsed by the United Nations—Food and Agriculture Organization (UN-FAO) Global Soil Partnership initiative as the specifications to follow in order to deliver a fine grid of soil properties. The *GlobalSoilMap* specifications do not prescribe the methods for generating the soil property predictions, because of diverse soil legacy data sources from various countries. A flow chart outlines the different approaches that can be applied to satisfy the specifications (Minasny & McBratney, 2010). Of particular importance to these specifications is the estimation of uncertainty which is also a major challenge of this project.

Several countries have already released products according to the *GlobalSoilMap* specifications (e.g., Adhikari et al. 2013, 2014; Akpa et al. 2014, 2016; Lacoste et al. 2016; Mulder et al. 2016a&b; Poggio & Gimona, 2014, 2016; Odgers et al. 2012; Hempel et al. 2014; Padarian et al. 2017; Viscarra Rossel et al. 2015) and the project is rejuvenating soil survey and mapping in many parts of the world (Arrouays et al. 2017b). Moreover, a global soil mapping working group was established under International Union of Soil Sciences (IUSS) to support the efforts of *GlobalSoilMap*.

In this paper, we review the *GlobalSoilMap* project according to the 5 dimensions of the Global Soil Security concept (McBratney et al. 2014; 2017), namely capability, condition, capital, connectivity and codification.

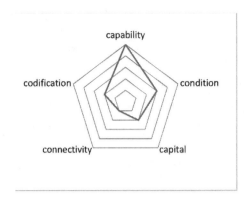

Figure 1. How the *GlobalSoilMap* project relates to the 5 dimensions of Global Soil Security.

that is the most connected to the *GlobalSoilMap* initiative. Capability and condition are strongly related.

2 GLOBALSOILMAP AND SOIL CAPABILITY

Capability refers to what functions a particular soil can perform (McBratney et al. 2017). As such, capability is strongly linked to intrinsic biophysical characteristics of soils. In the *GlobalSoilMap* specifications (*GlobalSoilMap*, 2015), the targeted information includes predicted values of selected key soil attributes at 6 standard depth intervals (0–5; 5–15; 15–30; 30–60; 60–100; and 100–200 cm), at a global scale, on a 3 arc-second support grid (approximately 90×90 m), along with their uncertainties. The key primary soil attributes include clay, silt and sand content, coarse fragments, pH, soil organic carbon (SOC), effective cation exchange capacity (ECEC) and soil depth to bedrock and effective root zone depth. Secondary key properties, mostly derived using pedo-transfer functions, include bulk density, and available water capacity. This list of attributes obviously constitutes the minimum dataset to assess which functions the soils can perform. In other words, *GlobalSoilMap* will provide a tool to assess the global soil functionalities and to quantify and map them at a high resolution across the world. It will also provide the necessary information to fill in the gaps in soil data and guide future efforts to increase the accuracy of soil property predictions. This is why *GlobalSoilMap* is strongly connected to soil capability assessment (Figure 1).

In some situations, however, this mandatory list of soil attributes may not be sufficient to assess the soil capability. This is why additional soil attributes may be added when necessary. These could be, for instance, attributes relevant to some limitations to performing some functions, such as sodicity, salinity, or water-logging. In summary, among the 5 dimensions of Global Soil Security, capability is the one

3 GLOBALSOILMAP AND SOIL CONDITION

Condition refers to the ability of soil to match its capability. Condition is a much more dynamic and sensible to management dimension compared to capability. Under this context, *GlobalSoilMap* product at a given date could be used as a baseline for assessing changes and monitoring trends of soil condition. Thus, *GlobalSoilMap* is not a static product, but is designed to improve, as new data become available. Therefore, a temporal series of *GlobalSoilMap* products may be considered as a global monitoring tool of soil condition. Indeed, collecting data at different times may be used to assess spatial-temporal soil changes and to perform multi-temporal data updates and queries. As an example, using legacy soil profiles data and land-use history maps, Stockmann et al. (2015) recently generated products following *GlobalSoilMap* specifications and incorporated a dynamic component. However, we recognize that a full monitoring of soil condition will require more data on soil attributes and management than those presently included in the *GlobalSoilMap* specifications. However, *GlobalSoilMap* can be seen as one of the building blocks of the global and local soil condition assessment.

4 GLOBALSOILMAP AND SOIL CAPITAL

Capital refers to the stock of materials and information, including physical structure and genetic information, contained in a soil (McBratney et al. 2017; Costanza, 1997). It can also refer to the value that is given to ecosystem goods and services provided by soil. The simplest examples relate to stocks, or to the potential of storage of elements. Soil organic carbon stock dynamics influence climate change either by mitigating or increasing greenhouse gases emissions. Carbon market and trading is a good example of an economic capital associated with soil. Available water storage capacity is another example where biophysical soil properties can infer an economical value of soil. *GlobalSoilMap* products enable a global and spatial assessment of carbon stocks, available water storage capacity and other capital related soil properties. However, the assessment of soil biophysical properties alone is not generally enough to define an economic value and many other considerations should be taken into account. In summary, by mapping and monitoring soil biophysical properties *GlobalSoilMap,* can establish the foundation to assess soil capital values and changes.

5 *GLOBALSOILMAP* AND SOIL CODIFICATION

Codification refers to policy and regulation applied to soil resources in order to limit soil degradation and to ensure that they are suitably and sustainably managed. The increasing soil degradation worsened under human driven pressures illustrates the current lack of such codification. The over-whelming conclusion from the State of the World's Soil Resources Report (FAO-ITPS, 2015; Montanarella et al. 2016) report is that the majority of the world's soil resources are in only fair, poor, or very poor condition. The current projections of soil condition may have catastrophic consequences (Amundson et al. 2015). Unless concerted actions are taken for improving soil codification, the situation will worsen and the consequences may be may be very pessimistic. Indeed, international treaties, conventions and legislations on soil are missing. The failure to launch a soil thematic strategy in Europe is an example of the difficulty to reach such agreements. The *GlobalSoilMap* initiative should improve our knowledge about the current state and trend of the soil condition. As such, it should provide guidance to policy makers to build a global strategy to fight soil degradation and to improve its sustainable management. To this respect, the endorsement of the *GlobalSoilMap* specifications by the Global Soil Partnership is an encouraging progress.

6 *GLOBALSOILMAP* AND SOIL CONNECTIVITY

This fifth dimension of soil security is probably indirectly linked to *GlobalSoilMap*. However, connectivity needs communication and education. One of the best way to communicate and educate on global issues related to soil condition is to provide maps showing those parts of the world where soil is "in good health" and those where it is severely degraded. Simple maps and figures, such as for instance maps of organic carbon stocks and simple 4perMille increase calculations (e.g., Minasny et al. 2017) may have a considerable impact on citizens and policy-makers alike.

7 CONCLUSIONS

Numerous countries and institutions have indicated their willingness to join the *GlobalSoilMap* initiative. At the end of 2016 a new working group of the International Union of Soil Sciences was established. In the near future, consistent high quality products will be updated when newly collected data become available. Ultimately, the *GlobalSoilMap* grid will provide suitable data to a wide variety of communities that makes decisions at various levels from local (field) to national scale and beyond. *GlobalSoilMap* is contributing to soil security assessment mainly through its contribution to capacity and condition assessment. Moreover, it may bring some inputs to the other dimensions of soil security. In particular, we believe that its contribution to soil capital assessment may re-inforce soil awareness to policy makers, soil managers and citizens and therefore increase the soil connectivity and codification dimensions.

REFERENCES

Adhikari K, Hartemink AE, Minasny B, Bou Kheir R, Greve MB, Greve MH. 2014. Digital Mapping of Soil Organic Carbon Contents and Stocks in Denmark. PLoS ONE; 9(8): e105519.

Adhikari K, Kheir RB, Greve MB, Bøcher PK, Malone BP, Minasny B, McBratney AB, Greve MH. 2013. High-Resolution 3-D Mapping of Soil Texture in Denmark. Soil Science Society America Journal; 77(3): 860–76.

Akpa SIC, Odeh IOA, Bishop TFA, Hartemink AE, Amapu IY. 2016. Total soil organic carbon and carbon sequestration potential in Nigeria. Geoderma; 271: 202–215.

Akpa SIC, Odeh IOA, Bishop TFA, Hartemink AE. 2014. Digital soil Mapping of soil particle-size fractions in Nigeria. Soil Science Society of America Journal; 78(6): 1953–1966.

Amundson R, Berhe AA, Hopmans JW, Olson C, Sztein AE, Sparks DL. 2015. Soil and human security in the 21st century. Science; 348(6235).

Arrouays D, Grundy MG, Hartemink AE, Hempel JW, Heuvelink GBM, Hong SY, Lagacherie P, Lelyk G, McBratney AB, McKenzie NJ, Mendonça-Santos MdL, Minasny B, Montanarella L, Odeh IOA., Sanchez PA, Thompson JA, Zhang G.-L. 2014. *GlobalSoilMap*: towards a fine-resolution global grid of soil properties. Advances in Agronomy; 125: 93–134.

Arrouays D, McKenzie NJ, Hempel JW, Richer de Forges AC, McBratney AB (eds). 2014. *GlobalSoilMap*: Basis of the global spatial soil information system. CRC Press Taylor & Francis Group; 478 p.

Arrouays, D, Johan Leenaars J.G.B.,, Anne C. Richer-de-Forges, A.C., et al. 2017. Soil legacy data rescue via *GlobalSoilMap* and other international and national initiatives. GeoRes J; 14: 1–19.

Arrouays D, Lagacherie P, Hartemink A. 2017. Digital soil mapping across the globe. Geoderma Regional; 9: 1–4.

Costanza, R 1997 'The value of the world's ecosystem services and natural capital' Nature, vol. 387, 253–260.

GlobalSoilMap Science Committee, 2015, http://www.globalsoilmap.net/specifications, last access 08/22/2016.

Grundy MJ, Viscarra Rossel RA, Searle RD, Wilson PL, Chen C, Gregory L.J. 2015. Soil and landscape grid of Australia. Soil Research; 53:835–844.

Hartemink AE, McBratney AB. 2008. A soil science renaissance. Geoderma; 148:123–129.

Hempel JW, Libohova Z, Thompson JA, Odgers NP, Smith CAS, Lelyk GW, Geraldo GEE. 2014. *GlobalSoilMap* North American Node progress. In: Arrouays D., McKenzie NJ, Hempel JW, Richer de Forges AC, McBratney AB (editors), *GlobalSoilMap*. Basis of the global spatial soil information system. 2014. Taylor & Francis, CRC Press; 2014. p. 41–45.

Koch A, McBratney A, Adams M, Field D, Hill R, Crawford J, Minasny B, Lal R, Abbott L, O'Donnell A, Angers D, Baldock J, Barbier E, Binkley D, Parton W, Wall DH, Bird M, Bouma J, Chenu C, Butler Flora C, Goulding K, Grunwald S, Hempel J, Jastrow J, Lehmann J, Lorenz K, Morgan CL, Rice CW, Whitehead D, Young I, Zimmermann M. 2013. Soil Security: Solving the Global Soil Crisis. Global Policy; 4(4): 434–41.

Lacoste M, Mulder VL, Richer-de-Forges AC, Martin MP, Arrouays D. 2016. Evaluating large-extent spatial modelling approaches: a case study for soil depth for France. Geoderma Regional; 7: 137–152.

Lelyk GW, MacMillan RA, Smith S, Daneshfar B. 2014. Spatial disaggregation of soil map polygons to estimate continuous soil property values at a resolution of 90 m for a pilot area in Manitoba, Canada. In: Arrouays D, McKenzie NJ, Hempel JW, Richer de Forges AC, McBratney AB, editors. *GlobalSoilMap*. Basis of the global soil information system, Oxon: Taylor & Francis, CRC press; p. 201–207.

McBratney A, Field DJ, Koch A. 2014. The dimensions of soil security. Geoderma; 213:203–213.

McBratney AB, Mendonça Santos MdL, Minasny B. 2003. On digital soil mapping. Geoderma. 2003; 117(1–2):3–52.

McBratney, AB, Field, DJ, Arrouays, D, Jarrett, LE. 2017d, 'Quantifying Capability: G*LobalSoilMap*' Global Soil Security, pp. 77–86.

McBratney, AB, Field, DJ, Morgan, CLS, Jarrett, LE. 2017a, 'Soil Security: A Rationale' Global Soil Security, pp. 3–14.

McBratney, AB, Field, DJ, Morgan, CLS, Jarrett, LE. 2017b, 'General Concepts of Valuing and Caring for Soil' Global Soil Security, pp. 101–108.

McBratney, AB, Field, DJ, Morgan, CLS, Jarrett, LE. 2017c, 'The Value of Soil's Contribution to Ecosystem Services' Global Soil Security, pp. 227–235.

Minasny B, McBratney AB. 2010. Methodologies for global soil mapping. In: Boettinger JL, Howell DW, Moore AC, Hartemink AE, Kineast-Brown S, (editors), Digital soil mapping: bridging research, envitonmental application, and operation, Springer Science +Business Media. P. 429–436.

Minasny, B., McBratney, A.B., 2016. Digital soil mapping: A brief history and some lessons. Geoderma 264, 301–311.

Minasny, B., Malone, B.P., McBratney, A.B., Angers, D.A., Arrouays, D., Chambers, A., Chaplot, V., Chen, Z.S., Cheng, K., Das, B.S., Field, D.J., Gimona, A, Hedley, C.B., Hong, S.Y., Mandal, B., Marchant, B.P., Martin, M., McConkey, B.G., Mulder, V.L., O'Rourke S., Richer-de-Forges A.C., Odeh, I., Padarian, J., Paustian, K., Pan, G., Poggio, L., Savin, I., Stolbovoy, V., Stockmann, U., Sulaeman, Y., Tsui, C-C., Vågen, T.-G., van Wesemael B., Winowiecki, L., 2017. Soil carbon 4 per mille. Geoderma; 292: 59–86.

Montanarella L, Pennock DJ, McKenzie NJ, Badraoui M, Chude V, Baptista I, Mamo T,Yemefack M, Singh Aulakh M, Yagi K, Young Hong S, P. Vijarnsorn P, Zhang G.-L, Arrouays D, Black H, Krasilnikov P, Sobocká J, Alegre J, Henriquez CR, Mendonça-Santos ML, Taboada M, Espinosa-Victoria D, AlShankiti A, AlaviPanah SK, Elsheikh EAE, Hempel JW, Camps Arbestain M, Nachtergaele F, Vargas R. 2016. World's soils are under threat. SOIL; 2:79–82.

Mulder VL, Lacoste M, Richer-de-Forges AC, Martin MP, Arrouays D. 2016. National versus global modelling the 3D distribution of soil organic carbon in mainland France. Geoderma; 263:16–34.

Mulder, V.L., Lacoste, M., Richer-de-Forges, AC, Arrouays, D. 2016. *GlobalSoilMap* France: High-resolution spatial modelling the soils of France up to two meter depth. 2016. Science of the Total Environment; 573: 1352–1369.

Odgers NP, Libohova Z, Thompson JA, 2012. Equal-area spline functions applied to a legacy soil database to create weighted-means maps of soil organic carbon at a continental scale. Geoderma; 189:153–163.

Padarian J, Minasny B, McBratney A.B. 2017. Chile and The Chilean Soil Grid: a contribution to *GlobalSoilMap*. Geoderma Regional; 9: 17–28.

Poggio L, Gimona A. 2017. 3D mapping of soil texture in Scotland. Geoderma Regional; 9: 5–16.

Poggio L, Gimona A. 2014. National scale 3D modelling of soil organic carbon stocks with uncertainty propagation. An example for Scotland. Geoderma; 232, 284–299.

Sanchez PA, Ahamed S, Carré F, Hartemink AE, Hempel JW, Huising J, Lagacherie P, McBratney AB, McKenzie NJ, Mendonça-Santos MdL, Minasny B,Montanarella L, Okoth P, Palm CA, Sachs JD, Shepherd KD, Vagen TG, Vanlauwe B, Walsh MG, Winowiecki LA, Zhang G.-L. 2009. Digital soil map of the world. Science; 325(5941):680–681.

Stockmann U, Padarian J, McBratney AB, Minasny B, de Brogniez D, Montanarella L, Hong SY, Rawlins BG, Field DJ. 2015. Global soil organic carbon assessment. Global Food Security; 6: 9–16.

Thompson JA, Nauman TW, Odgers NP, Libohova Z, Hempel JW, 2012. Harmonization of Legacy Soil Maps in North America: Status, Trends, and Implications for Digital Soil Mapping Efforts. In: Minasny B, Malone BP, McBratney AB (editors). Digital Soil Assessments and Beyond. 2012. CRC Press, Leiden, The Netherlands. p. 97–102.

Viscarra Rossel R, Chen C, Grundy M, Searle R, Clifford D, Campbell P. 2015. The Australian three-dimensional soil grid: Australia's contribution to the *GlobalSoilMap* project. Soil Research; 53(8): 845–64.

DSM for soil functions: A preliminary example using spatialized BBNs in Scotland

L. Poggio & A. Gimona
The James Hutton Institute, Aberdeen, UK

ABSTRACT: Soil plays a crucial role in the ecosystem functioning such as food production, capture and storage of water, carbon and nutrients. In this work we present an approach to spatially and jointly assess the multiple contributions of soil to the delivery of ecosystem services. We focused on the modelling of the impact of soil on sediment retention, carbon storage, storing and filtering of nutrients, habitat for soil organisms and water regulation. Simplified models were used for the single components. Spatialsed Bayesian Belief networks were used for the joint assessment and mapping of soil contribution to multiple ecosystem services. We integrated continuous 3D soil information derived from digital soil mapping approaches covering the whole of mainland Scotland, excluding the Northern Islands. Uncertainty was accounted for and propagated across the whole process. The Scottish test case highlights the differences in roles between mineral and organic soils. The results show the importance of the multi-functional analysis of the contribution of soils to the ecosystem service delivery.

1 INTRODUCTION

Soil can be considered as part of natural capital, providing a flow of goods and services. Erosion, decline in soil carbon and biodiversity can lead to soil degradation a serious global challenge for ecosystem sustainability. The contribution of soils to ecosystems services (ESS) requires appreciation beyond food production (Koch et al., 2013; McBratney et al., 2014). Soil resources need to be managed to provide multiple functions to support ESS delivery and to meet the United Nations Sustainable Development Goals (SDGs) (Bouma, 2014; Keesstra et al., 2016; Robinson et al., 2017). Soil resources support soil function, the delivery of ecosystem services and the UN SDGs (Keesstra et al., 2016; Robinson et al., 2017). It is therefore important to incorporate soils into the ecosystem services framework (Millennium Ecosystem Assessment, 2005), linking services with the multitude of functions provided. Only a few studies have linked soil properties to ecosystem services while most studies on the evaluation of ecosystem services lack completely the soil component or it is poorly defined and much generalized (Adhikari and Hartemink, 2016; Jonsson and Davidsdttir, 2016). Most of the mapping and modeling exercises used proxies to soil information, often in the form of land use and land cover data to produce spatially distributed biophysical parameter values needed for production function models (Adhikari and Hartemink, 2016). Monitoring and assessment information systems that can inform policy, regarding progress on achieving economic, social and environmental goals are important (Robinson et al., 2017). In this work we present an approach to spatially assess the multiple contributions of soil to the delivery of ecosystem services using Spatialised Bayesian Belief Networks (BBNs) (Gonzalez et al., 2016). Fig. 1 shows a schematic approach to the assessment of soil functions supporting multifunctional delivery of ESS. We focused on the modelling of the impact of soil on sediment retention, carbon stor-

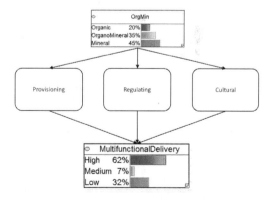

Figure 1. Schematic approach: BBN for one grid cell. OrgMin indicates the probability to have organic soils modelled with DSM approach defined and different for each pixel. The spatialiased BBN will use the pixel values.

age, storing and filtering of nutrients, habitat for soil organisms and water regulation.

2 MATERIALS AND METHODS

2.1 Soil information

The National Soil Inventory of Scotland (NSIS) contains descriptions, physical and chemical data (measured to a maximum depth of 1 metre) for profiles collected on a regular 10 km grid of sampled locations (Lilly et al., 2010). The following soil properties were mapped using the available data in NSIS:

1. depth of soil (censored at 100 cm) (Poggio and Gimona, 2017b)
2. pH measured in water (Poggio and Gimona, 2017b)
3. organic carbon (percent of content) (Poggio and Gimona, 2017b)
4. soil texture: as the relative percentage of sand (> 0.05 mm), silt (0.005–0.05 mm) and clay (< 0.005 mm), with the sum of the components always equals to the unit (Poggio and Gimona, 2017a)
5. Soil drainage classes: three drainage classes were derived from the definitions in the database (i.e. high, medium and low).
6. soil horizon types: the soil horizon were defined as organic or mineral (Lilly et al., 2010). The resulting map was used to differentiate the soil functions, as organic and mineral soils have different behaviour and characteristics.

2.2 Ancillary information

Ancillary information were used in the interpolation of soil properties and as additional information in the soil function modelling.

MODIS

A set of indices was derived from the Terra Moderate Resolution Imaging Spectro-radiometer (MODIS) 8 and 16 day composite products. The data were acquired from the NASA ftp website (ftp://e4ftl01u.ecs.nasa.gov/MOLT/) for 12 years between 2000 and 2012. The medians over 12 years (2000–2011) were used as covariates. The single images were restored to fill the cloud gaps using the approach described in Poggio et al. (2012). The indices selected were:

1. Enhanced Vegetation Index (EVI; Huete et al., 2002)
2. The Normalised Difference Water Index (NDWI; Gao, 1996)
3. Leaf Area Index (LAI; Knyazikhin et al., 1999)
4. Land Surface Temperature (LST; Wan, 1999)
5. Primary productivity (Running et al., 2004).

Sentinel-1

The Sentinel-1 (S1) mission provides data from a dual-polarization C-band Synthetic Aperture Radar (SAR) instrument. The data were preprocessed, prepared, mosaicked and downloaded from Google Earth Engine (Gorelick et al., 2017). The median of the images available between June 2015 and May 2016 was calculated for VV and VH polarisation. No further transformations were applied.

Sentinel-2

Sentinel-2 (S2) is a wide-swath, high-resolution, multi-spectral imaging mission supporting Copernicus Land Monitoring studies. The data were mosaicked and downloaded from Google Earth Engine (Gorelick et al., 2017). Several bands indices were calculated:

1. NDVI
2. NDWI (Gao, 1996) using band SWIR between 2100–2300.
3. Soil Moisture (SM) as ratio between NIR and blue bands
4. Soil Colour Index (SCI) as SCI = 3*NIR+Red – Green – 3 * Blue

Morphology

The Digital Elevation Model (DEM) used as a covariate in the fitted models was SRTM (Shuttle Radar Topography Mission), further processed to fill in no-data voids (Jarvis et al., 2006; Rodriguez et al., 2006): elevation, slope as the steepest slope angle, calculated using the D8 method (O'Callaghan and Mark, 1984), and topographic wetness index (TWI; Sorensen et al., 2006).

2.3 Interpolation of point data: Soil, climate and other features

An extension of the scorpan-kriging approach, i.e. hybrid geostatistical Generalized Additive Models (GAM Wood, 2006), combining GAM with kriging (Poggio and Gimona, 2014, 2017b) was used:

1. the fitting of a GAM to estimate the trend of the variable, using a 3D smoother with related covariates; and
2. kriging of GAM residuals as spatial component to account for local details, when residuals showed spatial auto-correlation.

The prediction grid had a resolution of 100 × 100 m for the lateral dimensions and 100 cm with 5 cm intervals for the vertical dimension. The results were then summarised for the standard GlobalSoilMap depths (Arrouays et al., 2014). The topsoil 0 to 15 cm was used in this example.

2.4 Soil functions modelling

Simplified models were used for the single components, in particular sediment retention, carbon storage, storing and filtering of nutrients, habitat for soil organisms and water regulation.

Erosion and sediment retention

A simplified version of the RUSLE approach (Wang et al., 2002) in a BBN framework was used for mineral soils. It was modified for organic soils to introduce specific rules (Lilly et al., 2002). The RUSLE factors were modified from Panagos et al. (2015):

R factor: interpolated from weather stations (Poggio et al, submitted).
K factor: interpolated from soil data.
C factor: integration of land use and remote sensing data.
P factor: interpolation of linear features and other management features.
LS factor: derived from the digital elevation model.

In this preliminary study, the sediment retention was calculated as the inverse of the soil erosion modifying the CPT table of the relevant node.

Carbon stocks

Carbon stocks were derived from the results described in Poggio and Gimona (2014). The interactions between soil erosion and carbon stocks were defined within the CPTs of the relevant nodes.

Suitability for biodivesity (soil organisms)

This initial approach was based on the description in Calzolari et al. (2016). CPTs were defined considering that that it is likely that soils rich in organic matter and not compacted are potentially capable to host a relatively higher biodiversity pool. Low and high values of pH were considered to imply low suitability for most microorganisms.

Soil filtering and buffering suitability

For this initial approach, soil filtering and buffering capacity was modelled following the suggestions of Calzolari et al. (2016). CEC, pH and OM, as humic acids and clay minerals have good buffering capabilities. Therefore the CPTs were adapted to reflect this from the input soil properties.

Soil and crop productivity

To model soil productivity for crops, the existing Land Capability for Agriculture (Bibby et al., 1982; Brown et al., 2011) was integrated with MODIS primary productivity (Running et al., 2004). The main soil properties used in the original LCA were organic matter, soil texture, drainage and pH, but they were not used directly in this study.

Water regulation

The water regulation potential was modelled using a measure of available water capacity (Poggio et al., 2010) and soil drainage. The CPTs were defined to illustrate the possible combinations of the two inputs properties.

2.5 Spatialised Bayesian Belief Networks (BBNs)

A BBN is a type of directed acyclic graph, where nodes are used to hold information on the variables in the model and their conditional interdependencies are represented by links or edges. Nodes that are not directly connected are assumed to be independent of each other. Within a BBN, each node has a defined set of states along with a conditional probability table (CPT), which defined for each child node state the probability of it occurring given all possible combinations of parent node states. Uncertainty is integral to Bayesian Belief Networks analysis. The BBN structure is flexible and BBNs can be updated easily as new information becomes available. See Jensen (2001) or Kjaerulff and Madsen (2013) for the theory of Bayesian Networks. The BBN building process followed a framework adapted from Marcot et al. (2006) and the Australian Department of the Environment, Water, Heritage and the Arts (DEWHA, 2010). Prior to using the spatial BBN software to produce the spatially explicit outputs, the GeNIes BBN software (Druzdzel, 1999) was used to develop the BBN framework (Landuyt et al., 2014). The probabilities from the conditional probability tables (CPTs) in each of the nodes are a product of different processes. The input node CPTs reproduce the data in the GIS layers. The CPTs of the nodes describing the soil functions and the multifunctional node were completed using expert opinion and literature search.

Spatialised BBNs were created using a combination of routines implemented in the R language (R Core Team, 2017; Scutari, 2010; Højsgaard, 2012). This enabled us to integrate the BBN constructed with GIS layers as well as non-spatial information, such as expert opinion (Gonzalez et al., 2016). The routines execute the BBN inference calculations for each of the raster grid cells of the input data. The outputs are raster maps where the pixel values correspond to the probabilities of a certain state for the target variable, i.e. the node for which the inference is calculated. Continuous variables were discretised using the quantiles approach.

3 RESULTS

3.1 BBN framework

The produced BBN framework is presented in Fig. 2. The spatial child nodes are divided in

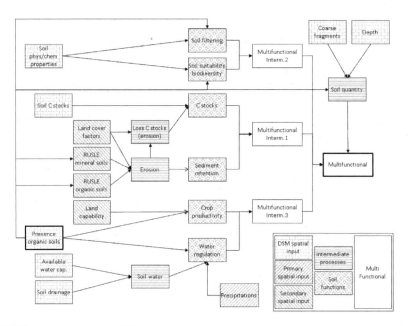

Figure 2. BBN framework: conceptual diagram.

primary and secondary. The primary GIS nodes are derived directly from climate and soil data with interpolation to obtain surfaces from points when necessary. Examples are the presence of organic soils, soil physical and chemical properties, precipitations. The secondary GIS are derived from multiple primary sources not included in the diagram. Examples are the RUSLE inputs, land cover factor, but also complex soil properties (i.e. carbon stocks and available water capacity) derived from some sort of pedo-transfer function. Primary and secondary child nodes feed into intermediate processes nodes, describing processes used for the definition of the soil functions, described in the next level of nodes. Finally soil functions nodes feed into the multifunctional analysis, with intermediate nodes necessary to make CPTs manageable.

3.2 Spatial outputs

Fig. 3 shows the maps of the probability of high delivery for the considered soil functions. The spatial patterns were consistent with the typical North-West South-East trend in Scotland. The patterns highlighted the differences between organic and mineral soils in Scotland. The models used so far to assess the soil functions were mainly addressing mineral soils. Additional work will be necessary to adapt and further develop for peat soils.

The spatial pattern obtained when taking into account multiple functions (Fig. 4) are markedly

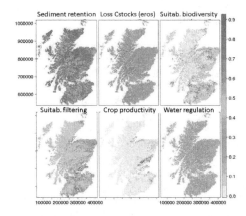

Figure 3. Probability of high delivery for single soil functions. Each pixel captures the probability to have high delivery of each function. Higher values indicate higher probability to have high delivery.

different from results obtained considering separate functions, highlighting the importance of considering multiple aspects at once when spatially assessing soil functions.

The results obtained with spatialised BBNs were expressed as mapped probabilities and reflect the propagation of uncertainty from input data to the final modelling stage and to the decision support. Spatialised BBNs could easily connect different models.

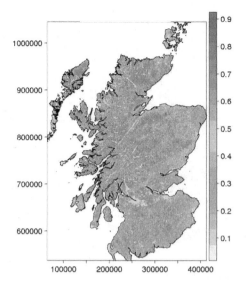

Figure 4. Probability of high multifunctional delivery. Each pixel captures the probability to have high multifunctional delivery. Higher values indicate higher probability to have high delivery.

4 DISCUSSION AND NEXT STEPS

In this preliminary study, we presented an approach to model, assess and map soil functions from soil properties maps. The results showed the importance of the multi-functional analysis of the different ecosystems and the contribution of soils to the ecosystem service delivery. It should be noticed that soil and vegetation components are interlinked and it is difficult to separate the contributions of each component to the function considered.

Multifunctionality was introduced as additional measure, in line with the Soil Security framework (e.g. McBratney et al., 2014). It is not, however, a direct indicator of soil quality. A soil with high multifunctional capability could have worst performances in the separate functions then a soil with low multifunctionality but very high performances in a single function (e.g. Carbon storage). Most soil quality indicators depend on the intended use (Vrscaj et al., 2008). The introductions of a multifunctional dimension is useful to assess capability for multiple uses. At the moment, the differences in potential among soils are not yet taken into account.

The models used for soil functions need to be further refined and adapted for high organic soils and the definition of the CPTs need to be refined for the considered models. A sensitivity analysis of the input data and CPTs would also be helpful for more detailed understanding and modelling.

The BBN approach proved useful to integrated knowledge about inter-relationships between soil properties and functions, enhancing the understanding of soil-landscape processes.

ACKNOWLEDGEMENTS

This work was funded by the Scottish Government's Rural and Environment Science and Analytical Services division. Many thanks to the team that sampled and analysed the soils and to the team that set up the database. MODIS data are distributed by the Land Processes Distributed Active Archive Centre (LP DAAC), located at the U.S. Geological Survey (USGS) Earth Resources Observation and Science (EROS) Center (lpdaac.usgs.gov). Sentinel data were available from the European Space Agency and Copernicus Service.

REFERENCES

Adhikari, K., Hartemink, A., 2016. Linking soils to ecosystem services a global review. Geoderma 262, 101–111.

Arrouays, D., Grundy, M.G., Hartemink, A.E., Hempel, J.W., Heuvelink, G.B., Hong, S.Y., Lagacherie, P., Lelyk, G., McBratney, A.B., McKenzie, N.J., d.L. Mendonca-Santos, M., Minasny, B., Montanarella, L., Odeh, I.O., Sanchez, P.A., Thompson, J.A., Zhang, G.-L., 2014. Chapter three globalsoilmap: Toward a fine-resolution global grid of soil properties. Vol. 125 of Advances in Agronomy. Academic Press, pp. 93–134.

Bibby, J., Douglas, H., Thomasson, A., Robertson, J., 1982. Land capability classification for agriculture. Tech. rep., The Soil Survey of Scotland – The Macaulay Institute for Soil Research, Aberdeen, UK.

Bouma, J., 2014. Soil science contributions towards sustainable development goals and their implementation: linking soil functions with ecosystem services. Journal of Plant Nutrition and Soil Science 177 (2), 111–120.

Bouma, J., Chenu, C., Flora, C.B., Goulding, K., Grunwald, S., Hempel, J., Jastrow, J., Lehmann, J., Lorenz, K., Morgan, C.L., Rice, C.W., Whitehead, D., Young, I., Zimmermann, M., 2013. Soil security: Solving the global soil crisis. lobal Policy 4 (4), 434–441.

Brown, I., Poggio, L., Gimona, A., Castellazzi, M., 2011. Climate change, drought risk and land capability for agriculture: implications for land-use in scotland. Regional Environmental Change 11 (3), 503–518.

Calzolari, C., Ungaro, F., Filippi, N., Guermandi, M., Malucelli, F., Marchi, N., Staffilani, F., Tarocco, P., 2016. A methodological framework to assess the multiple contributions of soils to ecosystem services delivery at regional scale. Geoderma 261, 190–203.

Druzdzel, M.J., 1999. SMILE: Structural Modeling, Inference, and Learning Engine and GeNIe: A Development Environment for Graphical Decision- Theoretic Models. In: Proceedings of the Sixteenth National Conference on Artificial Intelligence (AAAI99). Orlando, Florida (USA).

Gao, B., 1996. NDWI – a normalized difference water index for remote sensing of vegetation liquid water from space. Remote Sensing of Environment 58, 257–266.

Gonzalez, J., Luque, S., Poggio, L., Smith, R., Gimona, A., 2016. Spatial Bayesian Belief Networks as a planning decision tool for mapping ecosystem services trade-offs on multifunctional forested landscapes. Environmental Research 144, 15–26.

Gorelick, N., Hancher, M., Dixon, M., Ilyushchenko, S., Thau, D., Moore, R., 2017. Google earth engine: Planetary-scale geospatial analysis for everyone. Remote Sensing of Environment.

Højsgaard, S., 2012. Graphical independence networks with the gRain package for R. Journal of Statistical Software 46 (10), 1–26.

Huete, A., Didan, K., Miura, T., Rodriguez, E.P., Gao, X., Ferreira, L.G., 2002. Overview of the radiometric and biophysical performance of the MODIS vegetation indices. Remote Sensing of Environment 83 (1–2), 195–213.

Jarvis, A., Reuter, H., Nelson, A., Guevara, E., 2006. Hole-filled seamless SRTM data V3. Tech. rep., International Centre for Tropical Agriculture (CIAT).

Jensen, F., 2001.,Bayesian Networks and Decision Graphs. Springer.

Jonsson, J.O. G., Davidsdttir, B., 2016. Classification and valuation of soil ecosystem services. Agricultural Systems 145, 24–38.

Keesstra, S.D., Bouma, J., Wallinga, J., Tittonell, P., Smith, P., Cerd`a, A., Montanarella, L., Quinton, J.N., Pachepsky, Y., vanderPutten, W.H., Bardgett, R.D., Moolenaar, S., Mol, G., Jansen, B., Fresco, L.O., 2016. The significance of soils and soil science towards realization of the united nations sustainable development goals. SOIL 2 (2), 111–128.

Kjaerulff, U., Madsen, A., 2013. Bayesian Networks and Inuence Diagrams: A Guide to Construction and Analysis. Springer, Germany.

Knyazikhin, Y., Glassy, J., Privette, J., Tian, Y., Lotsch, A., Zhang, Y., Wang, Y., Morisette, J., Votava, P., Myneni, R., Nemani, R., Running, S., 1999. MODIS Leaf Area Index (LAI) and Fraction of Photosynthetically Active Radiation Absorbed by Vegetation (FPAR) Product (MOD15) Algorithm Theoretical Basis Document. Tech. rep., NASA Goddard Space Flight Center.

Koch, A., McBratney, A., Adams, M., Field, D., Hill, R., Crawford, J., Minasny, B., Lal, R., Abbott, L., O'Donnell, A., Angers, D., Baldock, J., Barbier, E., Binkley, D., Parton, W., Wall, D.H., Bird, M.,

Lilly, A., Bell, J., Hudson, G., Nolan, A., Towers, W., 2010. National Soil Inventory of Scotland 1 (NSIS1): site location, sampling and profile description. (1978–1998). Tech. rep., Macaulay Institute.

Lilly, A., Hudson, G., Birnie, R., Horne, P., 2002. The inherent geomorphological risk of soil erosion for overover flow in Scotland. Tech. rep., rep 183. Scottish Natural Heritage.

McBratney, A., Field, D., Koch, A., 2014. The dimensions of soil security. Geoderma 213, 203–213.

Millennium Ecosystem Assessment, 2005. Ecosystems and Human Well-being: Synthesis. Island Press, Washington, DC.

O'Callaghan, J.F., Mark, D.M., 1984. The extraction of drainage networks from digital elevation data. Computer Vision Graphics Image Processing 28, 323–344.

Panagos, P., Borrelli, P., Poesen, J., Ballabio, C., Lugato, E., Meusburger, K., Montanarella, L., Alewell, C., 2015. The new assessment of soil loss by water erosion in europe. Environmental Science & Policy 54, 438–447.

Poggio, L., Gimona, A., 2014. National scale 3D modelling of soil organic carbon stocks with uncertainty propagation - An example from Scotland. Geoderma 232–234, 284–299.

Poggio, L., Gimona, A., 2017a. 3D mapping of soil texture in Scotland. Geoderma Regional 9, 5–16.

Poggio, L., Gimona, A., 2017b. Assimilation of optical and radar remote sensing data in 3d mapping of soil properties over large areas. The Science of the Total Environment 579, 1094–1110.

Poggio, L., Gimona, A., Brown, I., 2012. Spatio- temporal MODIS EVI gap filling under cloud cover: an example in Scotland. ISPRS Journal of Photogrammetry and Remote Sensing 72, 56–72.

Poggio, L., Gimona, A., Brown, I., Castellazzi, M., 2010. Soil available water capacity interpolation and spatial uncertainty modelling at multiple geographical extents. Geoderma 160, 175–188.

R Core Team, 2017. R: A Language and Environment for Statistical Computing. R Foundation for Statistical Computing, Vienna, Austria, ISBN 3-900051-07-0. URL http://www.R-project.org/.

Robinson, D.A., Panagos, P., Borrelli, P., Jones, A., Montanarella, L., Tye, A., Obst, C.G., 2017. Soil natural capital in Europe; a framework for state and change assessment. Scientific reports 7 (Article number: 6706).

Rodriguez, E., Morris, C., Belz, J., Chapin, E., Martin, J., Daffer, W., Hensley, S., 2006. An assessment of the SRTM topographic products. Tech. Rep. JPL D-31639, NASA-Jet Propulsion Laboratory.

Running, S.R., Nemani, F.A., Heinsch, M., Zhao, M., Reeves, Hashimoto., H., 2004. A continuous satellite-derived measure of global terrestrial primary production. BioScience 54 (6), 47–560.

Scutari, M., 2010. Learning bayesian networks with the bnlearn r package. Journal of Statistical Software, Articles 35 (3), 1–22.

Sorensen, R., Zinko, U., Seibert, J., 2006. On the calculation of the topographic wetness index: evaluation of different methods based on field observations. Hydrology and Earth System Sciences 10, 101–112.

Vrscaj, B., Poggio, L., Ajmone Marsan, F., 2008. A method for soil environmental quality evaluation for management and planning in urban areas. Landscape and Urban Planning 88, 81–94.

Wan, Z., 1999. MODIS Land-Surface Temperature Algorithm Theoretical Basis Document (LST ATBD). Tech. rep., NASA.

Wang, G., Gertner, G., Singh, V., Shinkareva, S., Parysow, P., Anderson, A., JUL 15 2002. Spatial and temporal prediction and uncertainty of soil loss using the revised universal soil loss equation: a case study of the rainfall-runoff erosivity R factor. Ecological Modelling 153 (1–2), 143–155.

Wood, S., 2006. Generalized Additive Models: An Introduction with R. Chapman and Hall/CRC Press.

Soil systems—a soil survey approach to soil security

Zamir Libohova, Philip Schoeneberger, Doug Wysocki, Skye Wills & Cathy Seybold
United States Department of Agriculture, Natural Resources Conservation Service, National Soil Survey Center, USA

David Lindbo
United States Department of Agriculture, Natural Resources Conservation Service, Soil Science Division, USA

ABSTRACT: The concept of soil health has recently attracted the attention of many scientific and user communities worldwide including the US Department of Agriculture and specifically the Natural Resources Conservation Service (NRCS). This is understandable as maintaining healthy soils is important under the increased demands for more food and fiber and climate change uncertainties. At the present time, most soil health related efforts have focused activities at the individual field and farmstead scale. However, implementing a comprehensive program to effectively address soil health concerns over the long term and over wider areas (watersheds, countries, and regions) will require detailed soil data and information at management decision scales from field to watershed to regional levels and beyond. The soil survey concept based on soil landscape models has the potential to provide in-depth understanding of soil processes and properties and, more importantly, soil functions at these multiple scales. A single farm or field may span across various landscapes and vice versa. Each landscape and/or slope position within and between fields may function in unique ways with respect to plant growth and ecosystem functions. Such approach would also guide efforts for preserving and protecting the most productive soil landscapes while maintaining their health. We discuss the merits of approaching soil security and health from the soil landscape perspective. We argue that incorporating soil landscapes would broaden the scope and scale of soil security and has the potential to support and strengthen ongoing and future soil health initiatives.

1 INTRODUCTION

An international soil scientist panel outlined future research areas that includes soil pollution, erosion, soils and human health, urban soils, biodiversity, soil quality, climate change, environmental issues, food production, and water (1). Recent soil health initiatives in the United States including the Soil Renaissance (2) call for a soil health concept that includes not only soil fertility but also the biological and physical state of soil. This broader conception of the fundamental importance of soils is necessitated by climate uncertainty and the need to mitigate it by sequestering more carbon while sustaining high and stable yields in agricultural production. Because soils exist as part of a wider landscape, the societal needs and environmental concerns inherent in these topics require scientifically quantitative knowledge of spatial soil functions and an understanding of landscape-scale soil processes, the "soil system", (Fig. 1) in order to increase food security and other ecosystem services into the future. In this context, we focus the discussion on the soil-landscape concept as one of the major paradigms for many soil surveys worldwide (4, 5) and its usefulness as an integral part of soil systems and implications for soil health and security.

2 SOIL SYSTEM DEFINITION AND ADVANTAGES

A soil system is defined as a recurring group of soils that occupies the landscape from the inter-stream divide to the stream (3) and is characterized by similar soil parent materials, geomorphology, local relief, hydrologic connectivity, geographical extent, and climate. Soil systems may include many soil landscapes which are natural bodies that integrate ecosystem functions at an operational management human scale and highlight the key supporting role that soils play. The goal of soil systems framework is to collect state of the art data not only from soils but from other disciplines and areas of research in order to gain a comprehensive understanding of soil functions now and in the future (Fig. 1).

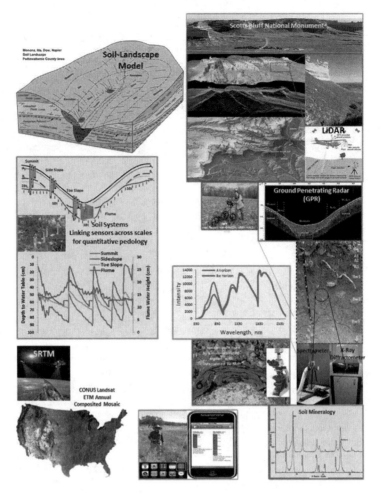

Figure 1. A Soil-Landscape model represents the building block of Soil Systems It bring the interactions of soil forming factors (climate, organisms, parent material, relief and time) (4, 5) into focus at a human scale; an operational scale. The integration of data from other disciplines with soil physical and biological processes in a soil system approach allows for a quantitative and dynamic soil functioning and understanding at scales relevant to those who actively manage, change, sustain and protect the land.

3 SOIL SYSTEM APPROACH

Soil System approach requires (i) knowledge about soil distribution; (ii) a quantitative characterization of soil functions; (iii) an in depth understanding and functioning of soils at the landscape scale; and (iv) an expansion beyond the current status and individual sites, fields and farms.

3.1 *Soil distribution*

For over a century many countries have conducted soil surveys and captured the knowledge about soil distribution by means of soil maps. The level of detail as determined by scale and demand has led to soil maps of various quality and extend (9). In addition, most of the first generation of soil maps have not been updated and brought to the digital platforms (9, 10). Several initiatives like *GlobalSoilMap* (9, 10, 11), Global Soil Partnership (12, 13), etc., have provided opportunities to rescue soil legacy data. Such efforts are essential in assessing the status of soil resources and planning future actions for improving, preserving and protecting soil resources in addition to quantitatively characterizing their functions and benefits. They also serve as a guidance for targeting major soil landscapes and soil systems for their in-depth characterization and hence preservation and protection.

3.2 Quantitative characterization of soil functions

Soil-landscape models (4, 5) have been the foundation of many soil surveys worldwide. However, almost exclusively, the soil survey products and information have been static and often qualitative and/or conceptual representations of soils and processes. Advances in data gathering and processing combined with the increased use of digital imaging and associated geospatial information for describing and mapping soils offers an opportunity for quantifying numerically soil properties and functions (5, 6) (Fig. 2). Soil systems represent the missing link that can bridge qualitative and quantitative soil information across multiple scales (6). For example, according to Soil Survey Geographic Database (SSURGO) water table depth for the month of March is greater than 200 cm (Fig. 2), yet according to water table monitoring instruments water table depth fluctuates between 0 and 100 cm during the same period.

3.3 In depth understanding and functioning of soils at the landscape scale

To be effectively implemented, soil systems requires that data gathering and analysis be conducted on representative soil landscapes that can be up scaled to larger areas by taking advantage of current accumulated soil knowledge and understanding of soil landscapes and their geographical distribution (Fig. 3). Traditionally, most of the soil fertility and liming tests have been based on surface soil samples and field or farm levels. This approach while suitable for planning short term management of inputs can

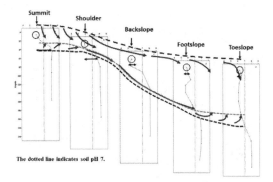

Figure 3. Soil pH variation with depth and slope position. Water movement and distribution as affected by soil-landscape influences the distribution of soil pH with depth (modified from Libohova et al., 2016).

result in long term disequilibrium and uneven distribution of soil resources. The surface soil pH (circles) varies only from 5.5 to 6.2 (Fig. 3). However, the soil pH depth profile increases rapidly with depth for the summit, shoulder and backslope, but less for footslope and toeslope positions. Depending on the type and depth of rooting systems the liming recommendations based only on surface samples may not account for the variability with depth which would affect the liming rate and distribution as well plant nutrient uptake. The soil pH concentrations and fluctuations are related to water movement throughout the soil column and across the landscape. The same can be inferred for other soil processes that can be better characterized and leveraged if they are understood on broader and natural contexts such as the entire soil thickness and/or soil landscape.

3.4 Expansion beyond the current status and individual sites, fields and farms

Many Soil Health and Quality Assessments focus mainly on the top 15 cm and might not be able to detect the continuous long-term decline of soil functions and services at the landscape scale (Fig. 4). The soil-landscapes in the US and worldwide continue to degrade (12) overall and in areas prone to prolonged winds such as mid-west US it could eventually lead to wind erosion and potentially trigger another "dust bowl" (14). The decrease of organic matter rich top soil layer has been also associated with decreases in soil organic matter (SOM) content (15). However, the decrease has not been uniform across soil landscapes leading in some cases to accumulation of more resistant SOM fractions in the lower parts of landscape and at the same time to depleted amounts of SOM on sloping areas but that are less resistant to mineralization (16) (Fig. 5).

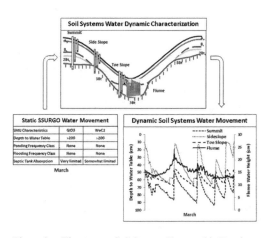

Figure 2. The current Soil Survey Geographic Database (SSURGO) provides map unit-centric, static information on soil water movement and related functions. Soil Systems would include dynamic soil water movement (temporal and spatial quantification) linked to affected soil processes and functions.

Figure 4. Public display in Iowa (central US—the "Corn Belt") that shows changes in the thickness of mollic surface over a 150 year period since European Settlement. The dark organic matter rich top soil layer thickness has decreased on average almost three times from 35 cm to15 cm.

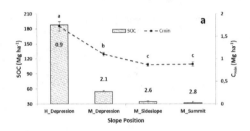

Figure 5. Total SOC and Cmin by Slope Position. Letters H and M represent Histic Epipedon and Mineral Soils, respectively. The amount of total SOM is higher for Histic epipedon soils in depressions compared to mineral epipedons in depressions and upslope positions. However, the amount of mineralized SOM (Cmin) is only 0.9% of total SOM compared to mineral soils on depression and summits with values ranging from 2.1 to 2.8%.

4 CONCLUSIONS

The soil systems and landscape-based approach can add a new dimension to soil health management and soil security. It combines existing knowledge with current demands for soil resources at practical scales where the majority of soil protection and conservation decisions are routinely being made. It offers an appropriate context to efforts for sustaining and improving the ability of soils to provide valuable ecosystem services.

REFERENCES

Arrouays D, Grundy MG, Hartemink AE, Hempel JW, Heuvelink GBM, Hong SY, et al. GlobalSoilMap: towards a fine-resolution global grid of soil properties. Adv. Agron 2014;125:93–134.

Arrouays D, McKenzie NJ, Hempel JW, Richer de Forges AC, McBratney AB, editors. GlobalSoilMap: basis of the global spatial soil information system. 1st ed.. CRC Press Taylor & Francis Group; 2014. p. 478.

Arrouays, D., Leenaars, G.B., Richer-de-Forges., A.C., 2017. Soil legacy data rescue via GlobalSoilMap and other international and national initiatives. GeoResJ 14: 1–19. http://dx.doi.org/10.1016/j.grj.2017.06.001.

Daniels, R.B., Buol, S.W., Kleiss, J., Ditzler, C.A. 1999. Soil Systems in North Carolina. Technical Bulletin # 314, Soil Science Dept. North Carolina State University, Raleigh, NC.

DeGloria, S.D., D.E. Beaudette, J.R. Irons, Z. Libohova, P.E. O'Neill, P.R. Owens, P.J. Schoeneberger, L.T. West, and D.A. Wysocki, 2014. Emergent Imaging and Geospatial Technologies for Soil Investigations. Photogrammetric Engineering & Remote Sensing 80(4):289–294.

Food and Agriculture Organization – Global Soil Partnership, 2015. Status of World's Soil Resources Main Report.

Food and Agriculture Organization – Global Soil Partnership, 2016. Global Soil partnership Pillar 4 Implementation Plan, Towards a Global Soil Information System. Developed by Pillar 4 Working Group and participants from the INSII Workshop (8–10 December, 2015).

Frontiers in Soil Science Research 2009: Report of a Workshop Steering Committee for Frontiers in Soil Science Research; National Research Council: ISBN: 0-309-13892-2, 80 pages. http://www.nap.edu/catalog/12666.html.

Hornbeck, R., 2012. The Enduring Impact of the American Dust Bowl: Short- and Long-Run Adjustments to Environmental Catastrophe. American Economic Review 2012, 102(4): 1477–1507. http://dx.doi=10.1257/aer.102.4.1477.

Jenny, H., 1941. Factors of Soil Formation: A System of Quantitative Pedology, McGraw Hill Book Company, New York, NY, 281 pp.

Libohova, Z., D.E. Stott, P.R. Owens, H.E. Winzeler, and S. Wills, 2014. Mineralizable Soil Organic Carbon Dynamics in Corn-Soybean Rotations in Glaciated Derived Landscapes of Northern Indiana. In A.E. Hartemink and K. McSweeney (eds.), Soil Carbon. Progress in Soil Science, DOI: 10.1007/978-3-319-04084-4_27, © Springer International Publishing, Switzerland.

Libohova, Z., Winzeler, H.E., Lee, B., Schoeneberger, P.J., Datta, J., Owens, P.R. 2016. Geomorphons: Landform and property predictions in a glacial moraine in Indiana landscapes. Catena 142:66–76. http://dx.doi.org/10.1016/j.catena.2016.01.002.

McBratney, A.B., M.L. Mendonsa-Santos, and B. Minasny, 2003. On digital soil mapping, Geoderma, 117:3–52.

Sanchez PA, Ahamed S, CarréF, Hartemink AE, Hempel JW, Huising J, et al. Digital soil map of the world. Science 2009;325 (5941): 680–1.

Soil Reconnaissance Strategic Plan, 2016. [Accessed on line October 2, 2016 at www.soilrenaissance.org].

Weismeier, M., Poeplau, Ch., Sierra, C.A., 2016. Projected loss of soil organic carbon in temperate agricultural soils in the 21st century: effects of climate change and carbon input trends. Scientific Reports 6:32525 | DOI: 10.1038/srep32525.

… Global Soil Security – Richer-de-Forges et al. (Eds)

Soil information system: The pathway to soil and food security in Haiti

Charles Kome, Paul Reich & Jessica Lene
United States Department of Agriculture, Natural Resources Conservation Service, World Soil Resources, USA

Zamir Libohova, Steve Monteith, Paul Finnell, Shawn McVey, Linda Scheffe, Susan Southard & Scarlet Bailey
United States Department of Agriculture, Natural Resources Conservation Service, National Soil Survey Center, USA

Tony Rolfes, Nathan Jones & Manuel Matos
United States Department of Agriculture, Natural Resources Conservation Service, USA

ABSTRACT: Land degradation from natural and anthropogenic causes are major threats to soil security that farmers and other land owners deal with on a regular basis. Managing and protecting soil resources requires information on the soil, how the soils were formed that enable us predict how soils respond to different types of disturbances, natural or anthropogenic. Soils and vegetation inventories are repositories of field, laboratory and derived data that can inform decision making in land use policy and planning at field, regional and national scales. Any sustainable agricultural system depends on the judicious use of land and water resources which require soil information. However, the lack of soil information may result in the mismanagement or abuse of soil resources, for example, excessive tillage, over irrigation or fertilization, inadequate soil cover, deforestation, soil harvesting of shallow soils, widespread urbanization of prime farmland have contributed to severe degradation of soil resources in Haiti as elsewhere in the world. Furthermore, lack of access to credit for agricultural inputs or needed conservation practices and the insecurities of some land tenure systems, gender inequity also contribute to neglect and land degradation. A pilot soil survey at 1:24,000 scale was conducted by the Haiti Ministry of Agriculture with technical assistance from USDA/NRCS on a 3,000-ha segment of Cul de Sac Valley as a proof of concept of the value and utility of soil information. Traditional and digital soil mapping techniques were used as part of a capacity building effort. A soil survey manuscript with spatial and tabular data on soil and vegetation inventories of the pilot area is now available as well as free on-line access to digital soil survey information (including portable devices like cell phones and tablets) within the pilot project area. In addition to the deliverables, Haitian scientists learned to generate soil interpretations from soil information, such as: soil functions related to water management, soil fertility landslides, flooding, and soil contamination, including soil erosion modeling with RUSLE2. Recognizing the value and utility of soil information technology the Haitian government is seeking funds to expand the survey to the entire country since the pilot project covered only 1% of the country. Soil information will inform land use policy and decision making to improve agricultural productivity based on the knowledge of the capabilities and limitations of each soil type thereby optimizing the use of soil types that are best suited for specific crops, cropping systems and other land uses in order to enhance soil and food security. Ready access to soil information has the potential to help protect Haiti's most valuable resources: its soils, its waters; its biodiversity and its people.

1 INTRODUCTION

Soil functions and services have been summarized as: a medium for plant growth that sustains and enhances net primary productivity (NPP) and agronomic yields to meet the demands (for food, feed, fiber, and fuel) of a growing world population; a foundation for buildings and civil structures; a source of raw materials for industry; a medium for storage and recycling of nutrients, water and carbon that mitigates climate change; a medium for disposing, denaturing and filtering pollutants, industrial and urban wastes; an archive of human and planetary history with provision of cultural and artistic values that contribute to our cultural heritage; a repository of germplasms and biodiversity; a medium that provides support and sustenance of a variety of landscapes and associated ecosystem services (Lal, 2009) on which our very existence depends. Soil is thus a valuable natural resource that is associated with or interacts with other life support systems such as air and water to provide various ecosystem services valued at about $33 trillion annually in 1994 US dollars (Costanza et al., 1997). Despite significant contributions to the human welfare, the value of soils is often discounted and taken for granted.

The combination of factors responsible for soil formation give rise to different soil types and characteristics with diverse physical, chemical, biological and engineering properties and a wide range of capabilities and limitations. A systematic inventory of soil properties (slope, texture, micro-relief, erosion potential, salinity, stoniness, particle size distribution, flooding potential and depth to bedrock or water table, infiltration, drainage, parent material, stoniness, pH, salinity etc.), shows the spatial variability which suggests different management requirements. Haiti has only a general soil survey at 1:250,000 scale with a few profile descriptions, measured data and very limited documentation.

Located between tropical latitudes 18 and 23° north Haiti's land area of 27,700 km^2 is primarily mountainous with 40% of all lands above 400 meters in elevation and mountain peaks as high as 2,684 meters sandwiching arable plains and valleys with about 63% of the landscapes having slopes greater than 20% and only 29% of landscapes with slopes less than 10%. Rainfall ranges from 300 mm in the northwest peninsular to 3 000 mm on the southwest mountains. Haiti is thus characterized by diverse moisture and temperature regimes which vary greatly with altitude. Extreme events such as hurricanes, droughts and floods are quite frequent, often with devastating consequences.

In the 1950s, agriculture provided employment for 80 percent of Haiti's labor force, contributed 50 percent of GDP, and 90 percent of exports. Poor land-use practices, drought, deforestation, hurricanes, population pressure, urban migration and increased soil erosion and degradation and crop failures have decreased agricultural sector contributions to less than 38 percent of employment and 26 percent of GDP. Haiti has faced significant food-security issues for several decades. A detailed accounting of the soil resources, their extent, capabilities and limitations would provide a blue print for designing long-term plans for protecting and managing soil resources to meet future soil and food security needs.

This pilot soil survey funded by the United States Agency for International Development (USAID) and implemented by MARNDR and USDA/NRCS, covered 3,000 hectares (7,500 acres) in the Plaine de-Cul-de-Sac. The purpose was to demonstrate the value of a soil information system. Based on feedback from USAID, government officials, farmers, and other stakeholders, MARNDR is proposing to develop a national soil survey program. This will significantly improve MARNDR's ability to improve agricultural productivity, natural resource management, conservation planning, environmental stewardship, and land use policy. This project will result in a soil information system tailored to meet the needs of the Ministry, individual farmers, land owners, and other vested entities and will be accessible to all. The objective of this project was to generate a detailed soil map for a pilot area in Haiti and build capacity to expand the soil survey to the entire nation.

2 MATERIALS AND METHODS

The soil survey was conducted based on a combination of traditional and digital soil mapping approaches. A preliminary digital soil map was developed using clustering algorithms of slope, curvature, and a topographic wetness index. Field observation sites were selected based on conditioned Latin Hypercube (cLHC) sampling methodology to assure representative sampling. Field observations, sampling protocols and laboratory analysis of physical and chemical properties and all data collection, were according USDA/NRCS guidelines. These information was used in map finishing. Several interpretative soil functions were generated based on soil distribution physical, chemical and biological properties.

Since soil erosion is a major problem in Haiti we prioritized erosion modeling to illustrate the value and utility of soil information system in conservation planning. In preparation, we interviewed farmers and conducted field observations to collect needed RUSLE2 input data and information. Climate data from the Thomazeau and Cornillon weather stations, soil properties from the pilot soil survey area, crop management calendars, as well as available records and expert opinion were used to develop RUSLE2 attributes. Sheet and rill erosion and soil quality trends based on the Soil Conditioning Index (SCI) (USDA/NRCS, 2003) for the pilot area were then estimated. Multiple runs were conducted for training purposes by comparing the effectiveness of alternative conservation practices on soil erosion and soil quality for the different soil types.

3 RESULTS AND DISCUSSIONS

The general soil map, detailed soil map and irrigation map are presented in Fig. 1. The general soil map showed a rich diversity of soil types within

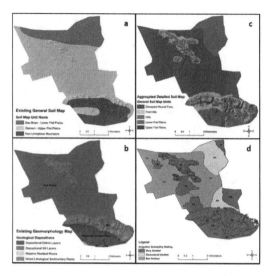

Figure 1. (a) Existing general map (b) Geomorphologic map (c) new detailed map (d) irrigation suitability soil map of the pilot area.

a very small geographical area. The pilot soil survey showed that only about 30% of the surveyed area was suitable for crop production with deep soils that have good structure and relatively high organic matter content (1–2%) for the 0–50 cm soil depth (Fig. 1). Best management and conservation practices, including sustainable land use will benefit the decision making especially under extreme weather conditions. The remaining 70% consists of soils that are limiting for crop production due to high levels of salts, carbonates, rock fragments and steep slopes. Soil interpretations were developed for water management, soil erosion control, soil fertility landslides, flooding, and soil contamination.

3.1 Soil erosion and soil quality modeling for sustainable agricultural systems in Haiti

Given the high cost of field measurements, models are normally used to predict soil erosion rates and then validated using limited field data. The 3000 ha USAID-funded pilot soil survey within the Cul-de-Sac plain provided the opportunity to use Haiti-specific soils, climate and crop management data to estimate soil erosion under various crop management scenarios using the RUSLE2 model.

Soil loss estimates ranged from less than 11 t ha^{-1} on the alluvial plains to over 300 t ha^{-1} with highly negative SCI scores for cultivated fields on very steep slopes. Ministry Leaders discussed recommendations for integrated sustainable systems that address soil, water, air, plant, animal, human,

and energy resource concerns for further evaluation in subsequent projects. Haitian scientists who previously used spreadsheets or handheld calculators to estimate soil loss using the USLE were impressed with the robust RUSLE2 tool and requested in-depth training to improve skills not only for water erosion prediction but for conservation planning. Skills learned during training and project activities contributed toward capacity building to meet Haiti's agenda for sustainable land use, soil security and water conservation and agricultural productivity.

In Haiti, as elsewhere in the world, the increasing demand for land for economic development, from population pressure and urban migration have triggered unprecedented land cover and land use changes with accelerated land degradation, resulting in soil loss, nutrient depletion, salinity problems, water scarcity, pollution, disruption of biogeochemical cycles, loss of biodiversity and soil productivity. Global annual soil loss of 75 billion tons cost about US$400 billion each year (Pimentel et al. 1995). FAO and World Bank soil loss estimates for Haiti from 1930 to 1999 (Jolly et al., 2009) showed an upward trend. Using regressional analysis this trend predicted annual soil loss estimates of about 50 million tonnes for 2011 with an approximate value of up to about 2.5 billion US dollars depending on the market (Fig. 2). These estimates do not include restoration and reclamation costs i.e. loading, transportation, spreading as well as offsite costs. Moreover, some adverse effects of land degradation are irreversible and the restoration or reclamation costs are expensive and most often prohibitive (Eswaran and Reich, 2001).

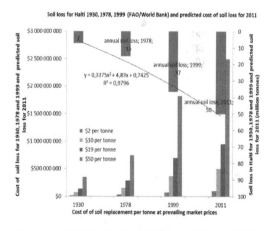

Figure 2. FAO/World Bank Soil Loss Estimates (1930, 1978, 1999, Jolly et al., 2009) and predicted soil loss for 2011 with cost estimates scenarios.

3.2 The economic benefits of soil survey

The economic value of soil or land is based on accessibility, suitability or limitation of the land for various uses. Standard land capability classes found in soil survey have been used extensively in land valuation and acquisition globally. Soil surveys have interpretations for crop yield, grazing, forestry, wildlife, recreation and camping sites potential. Other interpretations include military operations, pavements or road design, planning and zoning, effluent disposal, city planning, home and highway location and construction, cuts and embankments, infiltration systems, dikes, levees, ponds, agricultural drainage and irrigation, terraces, diversions and waterways, wood land productivity, rapid urban planning, sanitary landfills drain fields, parks, protected area, tourism, cites, etc. with widespread applications in city and regional planning.

Soil surveys are used for land valuation, and as a foundation for environmental, natural resource conservation, land use planning and as a guide for policy making, regulations and a wide range of other applications. Some examples include: proper placement of septic systems, erosion risk assessment, flood and ground water contamination potential, suitability and productivity for cultivated crops, trees, and grasses. Soil information is important to field engineers, tax commissioners, homeowners, land appraisers, community planners, farmers, ranchers, teachers and educators, realtors, fertilizer manufacturers, farm tool and equipment suppliers, foresters, zoning and planning commissions, investors and bankers, real estate dealers and home builders, watershed planning and environmental monitoring groups, highway departments, oil and gas companies, credit agencies, wildlife workers, and public officials.

All land-dependent applications and transactions require soil information for suitability and valuation purposes.

Soil surveys are also used as a tool for averting or reducing vulnerability to natural disasters such as hurricanes and floods by the documentation of the locations and extent of various soil risks and hazards to human life and property. Muckel et al., (2004) summarized soil-related concerns that can pose risks or hazards for many land uses (heavy metal contamination, compaction, corrosiveness drought, dust, earth wall collapse of soil pits and excavations, water and wind, erosion and sedimentation on construction sites and dams, stream bank erosion, expanding soils and shrink-swell potential, falling rocks, floods, frost action, gypsic soils, hydrocompactable soils, karst landscapes, landslides, liquefaction of soils during mudflows and earthquakes, radon potential, saline seeps, water-saturated soils etc.). Soil surveys thus serve as an invaluable tool for public officials, planners, developers, and others to consider soil information in land use decisions.

Land degradation from natural and or humaninduced causes affect the wealth of nations by reducing soil biodiversity, fertility, soil organic matter and moisture content, and hence soil health, affecting soil security and food security and sustainable livelihoods and is a direct consequence of population pressure (Eswaran and Beinroth, 1996) and lack of viable alternatives, inputs and resources and the absence of a conservation ethic. Simply stated, there can be no food security without soil security.

Soil loss and degradation reduce the nutrient, organic matter content, water holding capacity as well as soil and water quality and overall productivity. The economic toll from soil loss and degradation includes onsite losses such as soil displacement, (soil mass or volume with associated moisture, organic matter, nutrients and biodiversity) lost productivity, physical or structural damage as well as off-site losses (Fig. 3). Off-site losses include: siltation of dams, stream bank erosion, and high sediment loads in streams, water pollution etc., and the prohibitive cost of reclamation and rehabilitation.

In Haiti, significant investments are being directed toward stream bank stabilization, rehabilitation of dams and irrigation canals, roads, drainage ditches as well as municipal and industrial development, such as water and waste water treatment plants, construction of permanent settlements, schools, hospitals etc. These projects need soil survey information for proper design and implementation and post construction management and maintenance. Selected soil interpretations for the Cul-de-Sac pilot project are presented in Fig. 4.

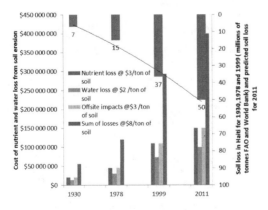

Figure 3. Soil loss for Haiti, 1930, 1978 and 1999 (FAO and World Bank, Jolly et al., 2009), predicted soil loss for 2011 with associated costs for nutrient loss, water loss, and offsite impacts.

The cost of production for soil surveys in the literature varies by country depending on the type or and scale of the soil survey; ranging from about $2.09 to $5 per acre for standard USDA/NRCS surveys (1:24,000 to 1:250, 000 scale) and up to $500 per acre for more detailed surveys (<1:12,000, Caldwell and Brown, 2003). The high benefits cost ratio ratios for soil surveys and soil information use from different countries (Table 1) consistently show that investments in soil surveys could be recovered within a few years of implementation of conservation practices, making a compelling case for investments in soil survey development for Haiti as elsewhere in the world

Theoretical frameworks for estimating the benefits of soil conservation using soil survey information include: reduced environmental impact or losses, reduced replacement costs, and reduced procurement for water, firewood and land and reduced relocation costs (Lew et al., 2001). Studies from different countries have reported high benefit /cost ratios when soil survey information was used in conservation planning (Table 1). The relatively low cost of an order 2 soil survey (1:24000 scale) is quickly offset by the benefits (financial and otherwise) that come from appropriate land use: using the right soil for the right purpose, or taking necessary measures to include improvements to compensate for site and soil-specific limitations during project design to avoid failures or avert other risks (Brown, 1991).

In the case of Haiti, the partial replacement costs for water, nutrients alone exceed the cost of soil survey development by several orders of magnitude even for the low 1930soil loss estimates (Figs. 2 and 3). Replacing half an inch of top soil may cost as much as $1800 per acre or $4500 per hectare in some parts of the US.

Assessing the value of soil information using decision analysis, Giasson et al., 2000 estimated a hypothetical economic benefit from improved management based on soil survey information at $17.14 per hectare compared to the cost of conducting a soil survey of $2.09 per hectare. The cost of their soil survey was paid for 7 times over in the first year. Using 14 different categories of soil conservation benefits to estimate the cost of erosion per ton of soil (Hansen and Ribaudo, 2008) concluded that most models capture only lower bound estimates of the economic value of soil and economic benefits of soil conservation.

Haiti is presently going through a significant phase in its development, a process that not only requires but will benefit from a well-structured soil information system that addresses relevant current and future land use interpretations. The foregoing discussion illustrates that economic returns derived from local, regional or national investments in soil surveys go beyond the ornamental value of beautiful soil maps With this information in hand, decision-makers will be able to plan with confidence, knowing that development projects are based on sound science and in doing so, reduce or avert risks, wasted time, money, effort and unnecessary environmental degradation and associated human and environmental health risks. Investments in soil survey provide an inventory of personal and national natural capital whose economic value is priceless. A soil survey for Haiti will help delineate areas suitable for different types of agriculture and other uses to mitigate vulnerability to drastic weather patterns in order to achieve lasting improvements in soil, air and water quality for its watersheds which are invariably linked to food security and livelihoods of its people.

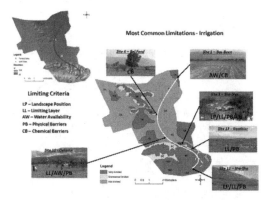

Figure 4. Selected soil interpretations to demonstrate the value and utility of soil information.

Table 1. Benefit/cost ratios for selected applications of soil information from soil survey for conservation planning by country.

Country	Benefit/cost	Source
Brazil	7:1	Giasson et al., 2000. Soil Sci. Vol. 165: 12:971–978
Italy	6:1	Fais and Bonati (http://www.inea.it/cartgrapfia.html)
New Zealand	6:1 and 13:1	Carrick et al. Landcare, New Zealand
US	5:1, 13:1	Fletcher et al., 2009. West Virginia University
US	45:1 (range/woodland)	Caldwell and Brown. http://edis.ifas.ufl.edu/ss160
US	61:1	(cropland) Caldwell and Brown. http://edis.ifas.ufl.edu/ss160
US	123:1 (urban areas)	Caldwell and Brown. http://edis.ifas.ufl.edu/ss160

REFERENCES

Bonati, G. & Fais, A. 1996. Factors affecting the uptake of information technologies in extension services in South Italy: the SILA experience. Proceedings ICCTA '96. Wageningen (NL) '96.

Brown, R.E. 1991. http://edis.ifas.ufl.edu/ss160.

Caldwell, R.E. and R.B. Brown. 1990. The Nature and Use of a Soil Survey. Fact Sheet SL-11, a series of the Soil and Water Science Department, Florida Cooperative Extension Service, Institute of Food and Agricultural Sciences, University of Florida.

Carrick, S., V. Éva-Terézia V. and A. Hewitt. Economic value of improved soil natural capital assessment: a case study on nitrogen leaching. Landcare Research, PO Box 40, Lincoln 7640, New Zealand, Email: carricks@landcareresearch.co.nz.

Costanza, R.R. D'Arge, R. De Droot, S. Farber, M. Grasso, B. Hannon, K. Limburg, S. Naeem, R.V. Oneill, J. Paruelo, R. Raskin, P. Sutton, and M. Van Den Belt (1997). The value of the world's ecosystem services and natural capital. Nature 387:253–260.

Eswaran, H. and Beinroth, F.H. 1996. Land degradation: issues and challenges. Abstracts. International Land Degradation Conference, Adana, Turkey (10–14 June 1996) University of Cukurova, Department of Soil Science, Adana, Turkey.

Eswaran, H., Lal, R. and Reich, P.F. 2001. Land Degradation: An Overview. Responses to Land Degradation. Proceedings of the 2nd International Conference on Land Degradation and Desertification, Khon Kaen. Oxford Press, New Delhi.

Fais, A. and G. Bonati; http://www.inea.it/cartgrapfia.html.

Food and Agricultural Organization of the United Nations, (FAO) Report. Land Resources information systems in the Caribbean. http://www.fao.org/docrep/004/Y1717E/y1717e13.htm.

Giasson, E., C. van Es, A. van Wambeke and R.B. Bryant. 2000. Assessing the economic value of soil information using decision analysis techniques. Soil Science 165: 971–978.

Hansen, L. and M. Ribaudo. 2008. Economic measures of soil conservation benefits. Regional values for policy assessment. USDA/ERS Technical Bulletin No. 1922.

Jolly, C.M. D Shannon, M. Bannister, G. Flaurentin, D. John, A. Binns, and P. Lindo. 2006. Income efficiency of soil conservation techniques in Haiti. CASES 26th West Indies Agricultural Economic Conference, Puerto Rico.

Lal, R. Ten tenets of sustainable soil management. 2009. Journal of Soil and Water Conservation vol. 64 no. 1.

Lew, D. K, D.M Larson, H. Suenaga, R. DeSousa. 2001. The Beneficial use of values database. University of California, Davis, Department of Agricultural Economics, April. http://buvd.ucdavis.edu/buvd.web.pdf.

McBratney, A.B., M.L. Mendonsa-Santos, and B. Minasny 2003. On digital soil mapping, Geoderma, 117:3–52.

Minasny M. and McBratney A.B. 2006. A conditioned Latin hypercube method for sampling in the presence of ancillary information. Computers & Geosciences 32: 1378–1388. doi:10.1016/j.cageo.2005.12.009.

Muckel et al (20 Muckel, Gary B. (ed.).Understanding Soil Risks and Hazards. USDA-NRCS. National Survey Center. Lincoln, Nebraska. 2004.

Pimentel, D., C. Harvey, C., P. Resosudarmo, K. Sinclair, D. Kurz, M. McNair, S. Crist, L. Shpritz, L. Fitton, R. Saffouri, and R. Blair. 1995. Environmental and economic costs of soil erosion. Science, 267, 1117–1123.

United States International Development Agency, United States Department of Agriculture, Natural Resources Conservation Service, and Haiti Ministry of Agriculture. 2014. Soil survey of Cul de Sac, Haiti. http://soils.usda.gov/survey/printed_surveys. Note 16. Washington, DC: USDA Natural Resources Conservation Service. http://soils.usda.gov/sqi/.

Soil security and practitioners

Overview of tillage practices and correlations with other practices in France: An analysis of the agreste survey (2011)

N. Cavan
GIS GC HP2E, Paris, France

J. Labreuche & A. Wissocq
ARVALIS – Institut du Végétal, Boigneville, France

F. Angevin
INRA, Unité Eco-Innov, Thiverval-Grignon, France

I. Cousin
INRA, UR 0272 SOLS, Orléans, France

ABSTRACT: The tillage practices modify the soil condition (structure, porosity, etc.) and are also correlated to other practices of the crop sequence (such as cover crop use) modifying soil condition too. Therefore, the tillage practices chosen by farmers may be correlated to the type of soil and its capability. In order to study correlations between agricultural practices and tillage on farmers' fields, we used the "Agreste – Enquête Pratiques culturales 2011" survey, led by the French Ministry of Agriculture. We demonstrated that calcareous clay soils are correlated to greater use of reduced tillage but there was no evidence of a correlation between ploughing and poorly structured soils. No-ploughing tillage is correlated to greater use of herbicides for weed control, smaller use of other inputs and smaller yield. Cover crops are more diversified and destructed later in NPT fields (three weeks), although cover crop does not cover more surfaces during intercrop period.

1 INTRODUCTION

Tillage practices have many impacts on cropping systems, as they modify soil condition: structure (fragmentation, mixing, and compaction), air and water circulation, organic matter (OM) repartition, etc. (Roger-Estrade et al., 2014). Those operations have consequences on seeding quality (Labreuche et al., 2014), carbon and nitrogen mineralization (Mary et al., 2014), and contribute to weed control by burying weed seeds or destructing germinated seeds (Colbach and Vacher, 2014).

During a crop sequence, different tillage operations are performed, forming a tillage method (TM). Roger-Estrade et al. (2014) used the deepest tillage operation to characterize a TM, as it describes dilution of nutrients and organic matter in the soil profile, deep fragmentation probability, and gives an indication on the TM cost. They separated ploughing (defined by tilled layer inversion) from no-ploughing tillage (NPT), the latter including lots of operations, from non-turning implement to no-till.

In France, NPT use was correlated to a small yield loss compared to ploughing by Labreuche et al. (2007). However, there was great variability between crops, and a part of this variability is assumed to be due to pedoclimatic conditions. According to Soane et al. (2012) metanalysis on European no-till experiments, calcareous clay soils, with their self-mulching capacity, are considered to be the most suitable soils for no-till practice, contrary to poorly structured soils (loamy textures and/or poor OM content) or hydromorphic soils where NPT use could cause more yield loss compared to ploughing.

Using a French national survey on farmers' practices, the aim of this paper is to i) investigate correlations between tillage practices and soil capability; ii) identify correlations between tillage and other practices in farmers' fields.

2 MATERIAL AND METHOD

We have used the survey "Agreste – Enquête Pratiques culturales 2011" from the French Ministry of Agriculture. Almost every five years since 1986, some farmers (around 20 000 in 2011) have to describe all the cultural practices they made one of their field during the crop sequence.

2.1 Available data

Among the available data in the survey, we used the following ones: i) localization; ii) sowed crop in the 2010–11 cultural campaign, including wheat, maize, rapeseed, barley, sunflower, sugar beet, potato, peas; and crops sowed in the five previous years (2006–2010); iii) all tillage operations for the 2010–11 campaign, and summarized data for the 2006–2010 period, *i.e.* ploughing or NPT; iv) data for all agricultural practices on the 2010–11 campaign: fertilization, use of plant protection products, cover crop management, crop performances. The "Agreste – Enquête Pratiques Culturales 2011" survey contains data on soils, based on BaseSol[1], a database of French agronomical soil types defined by ARVALIS – Institut du végétal. A type of soil from the BaseSol database is declared for each surveyed field, described by a common name and five discrete variables: texture of the tilled horizon, depth, proportion of coarse fragments, hydromorphy and calcareous content (Fig. 1).

2.2 Conditions for data use

Some rules apply on data to respect farmers' privacy: a group of surveyed fields can only be described if i) it contains at least 3 fields; ii) no field represents 85% or more of the surveyed area of the group. Moreover, Agreste applied a statistical treatment on field surfaces and recommended to only use groups of minimum 30 fields (Agreste, 2014). Studying tillage correlations with other practices and crop performances implies to compare ploughed fields and NPT fields in a similar agronomic and pedoclimatic context. Therefore, typologies were created to diminish the number of possible contexts, and maximize available data for each analysis.

2.3 Description of tillage operations

Tillage methods (TM) were used to describe tillage practices during the 2010–11 campaign based on Roger-Estrade et al. (2014) typology (Fig. 2). Correlations between tillage and other practices were investigated on this campaign, by using a simplified 2-classes typology for TM: ploughing or NPT.

Tillage strategies (TS) are defined by summarized data on the 2006–10 period (ploughing/NPT – §2.1) and by data on the 2010–11 crop sequence. It provided a new variable: the number of years with ploughing (0 to 6 years), used to define three main TS: all-ploughing strategy (6 years ploughing), all-NPT strategy (0 year) and alternate ploughing-NPT (1–5 years).

2.4 Definition of agronomical typologies

Three variables were used to describe the agronomical context: surveyed crop, previous crop (2009–10), and succession of crops on the 2006–2011 period, defining a crop rotation. For the surveyed crop, those described in the survey were used.

For previous crops, 10 types of previous crops were described, based on harvesting period, crop residue production and their management techniques: cereal crop, corn and silage corn, sugar beet, potato, rapeseed, sunflower, protein crops, pasture and other previous crops.

For crop rotations, three criteria were used to define types of crop rotation: i) amount of cereal crops sowed (from 0 to 6 in six years); ii) amount of spring crops sowed (from 0 to 6); iii) presence of grassland (at least 2 years or more). Six types of rotation were defined: fall cereal crops; spring cereal crops, fall & spring cereal crop, fall cereal & dicotyledonous (C&D) crops, fall & spring and C&D crops, and crop rotations with mainly grasslands.

2.5 Definition of pedoclimatic typologies

The location of surveyed fields was used to estimate a climatic context, using production areas defined by ARVALIS – Institut du végétal (Fig. 3).

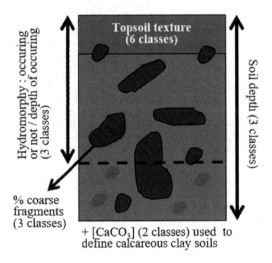

Figure 1. Variables used to describe the type of soils in the BaseSol database from ARVALIS – Institut du végétal.

---

1. https://plateforme.api-agro.fr/explore/dataset/referentiel-sol-arvalis-2017/: last visit 2017/09/27.

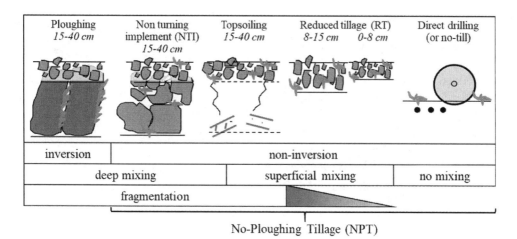

Figure 2. Tillage methods typology defined by inversion, mixing, fragmentation and tillage depth (indicated in italic on the figure). This typology is adapted from Roger-Estrade et al. (2014) to fit Agreste's survey data. Strip till is not represented as only a few fields with this tillage practice were surveyed; reduced tillage is described in two classes depending of the depth of work: 8–15 cm and 0–8 cm.

Figure 3. Production areas. Source: ARVALIS – Institut du végétal. These areas are defined by regrouping French administrative regions, based on crops and type of farms. It was used as a climate context proxy.

Preliminary results (Cavan et al., submitted) indicated that surfaces for each TM are correlated to the texture of the tilled horizon mainly (6 classes), and less correlated to hydromorphy. Moreover, there is no significant correlation between soils coarse fragments content and TM.

Soane et al. (2012) demonstrated that calcareous clay soils were more suitable for systems using NPT because of their high self-mulching capability. In order to evaluate this proposition on French farmers systems, calcareous clay soils were defined as a soil-type, based on the common name of the type of soil in the database and on the calcareous content.

Due to conditions for data use (cf §2.2), some soil textures were gathered, giving a 4-class variable: clay, clayey silt, silt & loamy silt, and silty loam & loam. For soils defined by silt & loamy silt texture, there was enough surveyed fields to define two soil-types: hydromorphic and non-hydromorphic.

Six soil-types are finally defined: i) clay; ii) calcareous clay soil; iii) silty clay; iv) non-hydromorphic silt & loamy silt; v) hydromorphic silt & loamy silt; vi) loam & silty loam. Distribution of production areas surfaces by soil-type is shown in Figure 4. Every soil-type is present in each production area, with quite different distributions. Soil-type distribution is quite balanced in production area 1 (Fig. 3); calcareous clay soil-types (respectively silt & loamy-silt soil-types) are overrepresented in production areas 4 and 7 (respectively areas 2 and 3).

Every factor, except surveyed crop, was described by typologies to simplify survey data and prevent data loss by application of restrictions described earlier. With these typologies, 52 contexts were identified where ploughed and NPT fields can be compared.

2.6 *Statistical analysis*

The results of this overview of tillage practices in France are based mostly on Principal Components Analysis (PCA), associated with Ascending Hierarchical Clustering (AHC), using R software (package FactoMineR, v. 1.28, 09/26/14). We used

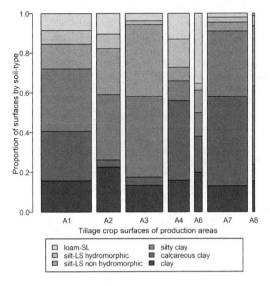

Figure 4. Tillage crop surfaces of production areas and distribution by soil-type. Source: Agreste – Enquête Pratiques culturales 2011. Production area 5 is not represented as only too few fields were surveyed there. A1: Production area 1. loam-SL: loam & silty-loam. silt-SL: silt & loamy-silt.

Table 1. Available data for each type of results. Source: Agreste – Enquête Pratiques culturales 2011.

Results	Factors	Groups	Fields	Data loss
TM distribution	Surveyed crop Previous crop Prod. Area Soil type	194	11718	44%
TS distribution	Crop rotation Prod. Area Soil type	143	14735	29%
Other practices*	All factors	52	3157	85%

*Correlations between tillage and other practices.

the proportion of surfaces where each TM or TS is used as quantitative variables, and the previously defined typologies as qualitative factors.

Correlations between PCA components and qualitative factors were described using Fishers' F statistics. Concerning the AHC, two statistical tests were used to describe each cluster: i) a t test of Student to identify when the average proportion of surfaces covered by a TM or a TS in the cluster is higher/smaller than the average for all groups; ii) a customized test (Lê et al., 2008) to describe under or over-represented types of the factors in the cluster.

In order to study correlations between tillage practices (ploughing or NPT in 2010–11 campaign) and other practices, groups of surveyed fields with at least 26 individuals, and 6 fields minimum for each type of tillage (ploughing and NPT) were used. For each quantitative variable describing farming practices, we calculated averages for ploughed fields (m_{plough}) and for NPT fields (m_{NPT}) as well as the difference between them (d): $d = m_{plough} - m_{NPT}$. A t test of Student was performed on the average of d for all the groups to determine if it is significantly different from 0. We also used PCA and AHC to compare cultural practices correlations with tillage practices and all the other factors defined earlier.

Although simplified typologies were created to prevent it, Table 1 shows data loss for each analysis.

3 RESULTS

3.1 Tillage methods distribution: A function of crop, previous crop, production area and soil-type

TM distribution on tillage crop surfaces was studied with 194 groups of fields (11718 fields) (Table 1). TM surfaces are mainly correlated to the surveyed crop (52% of the PCA first component variability).

The first cluster of the AHC is characterized by over-represented (§ 2.6) spring crops (corn and sunflower) and more surfaces under ploughing whereas fall crops (rapeseed, durum wheat, wheat and winter pea) are over-represented in cluster 2, with more surfaces under reduced tillage (RT) (cf. Fig. 5). This is coherent with general results of the survey, with 40 to 58% of surfaces of fall crops under NPT, compared to 15–30% for all spring crops (Agreste, 2014). Moreover, calcareous clay soils are also over-represented in cluster 2.

Cluster 3, composed of 24 groups, was divided in two parts (by another PCA + AHC). First part is composed of 14 groups, characterized by a larger use of RT (27% of surfaces compared to 14% on average), mainly fall sown crops on soil-types containing 20% clay or more. In the other part (10 groups), 24% of surfaces are under deep NPT (NIT and topsoiling), compared to 8% on average. Five of these groups are from production area 4 (South-West of France) and among them three are described by the hydromorphic silt & loamy silt soil-type.

Cluster 4 is formed by the only 4 groups of fields with surfaces under no till: there is not enough groups to describe a context associated to this TM.

3.2 Tillage strategies distribution: A function of crop rotation, production area and soil-type

TS distribution on tillage crop surfaces was studied with 125 groups of fields (14735 fields)

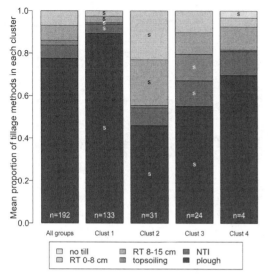

Figure 5. Tillage methods distribution in each cluster (clust) of the AHC. Source: Agreste – Enquête Pratiques culturales 2011. n is the number of field groups in the cluster. The average of field groups' surface distribution is represented for all groups and for each cluster. s: average proportion of a TM is significantly higher/lower in the considered cluster compared to average proportion for all groups.

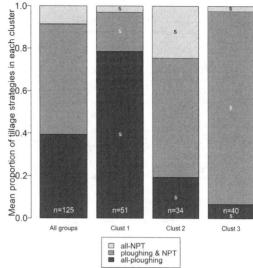

Figure 6. Tillage strategies distribution in each cluster (clust) of the AHC. Source: Agreste – Enquête Pratiques culturales 2011. n is the number of field groups in the cluster. The average of field groups' surface distribution is represented for all groups and for each cluster. s: average proportion of a TS is significantly higher/lower in the considered cluster compared to average proportion for all groups.

(Table 1). TS are mainly correlated to crop rotation: 81% of PCA first component variability is explained by this factor. Cereal-based rotations are correlated to the all-ploughing strategy. Rotations with fall crops and cereal & dicotyledonous crops are more correlated to all-NPT strategies, or alternate ploughing & NPT, especially for fall crops. The PCA 2nd component describes correlations between all-NPT strategy and area 7 (North-East of France) & calcareous clay soils. In this area, calcareous clay soils represent 45% of tillage crop surfaces, compared to 24% on national average.

Concerning the AHC, only crop rotation and soil-type were correlated to the three identified clusters. TS surfaces in each cluster are presented in Figure 6. The first cluster (all-ploughing strategy) contains groups with cereal-based rotations (mainly with rotations combining fall and spring crops, and a quarter of spring crop rotations – e. g. corn monoculture). In the second cluster, all-NPT strategy is twice as important as in average (25% of surfaces – Fig. 5); rotations with cereal and dicotyledonous crops and calcareous clay soils are over-represented in this cluster. The third cluster contains mainly groups with grassland rotations (maximum 2 tillage crops in six years): hence, ploughing & NPT strategy represents 90% of surfaces of this cluster.

3.3 Correlations between tillage practices and all other practices of the crop sequence (2010–11)

Significant differences between ploughed and NPT fields for the full crop sequence (2010–11) are shown on the Figure 7. In NPT fields, there is on average +0.7 tillage operation than in ploughed fields: farmers generally compensate ploughing by another type of tillage operation. Although there is no difference on residues management and land cover by cover crops, species seems to be more diversified in NPT fields than in ploughed fields (where mostly mustard is sown). Cover crop destruction is delayed by almost 3 weeks on NPT fields, and is generally done in a chemical way, whereas mechanical destruction is more common on ploughed fields. Concerning inputs (fertilizers and plant protection products), they are usually higher in ploughed fields, as crop yield (+ 2.7 q/ha). To describe pesticides use, we used the Treatment Frequency Index (TFI): the number of registered doses of pesticides used per hectare for one cropping season. Contrary to others inputs, the use of herbicides is greater in NPT fields: + 0.4 TFI (average of 1.5 TFI in ploughed fields), mostly for weed control purposes (+0.37 TFI).

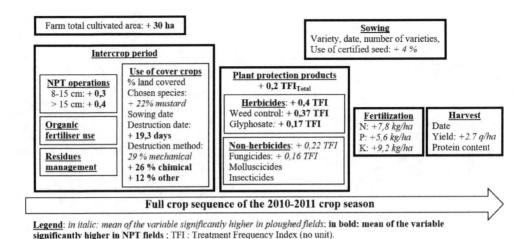

Figure 7. Correlations between tillage practices (ploughing or NPT) and other cultural practices of the 2010–11 crop sequence, for 52 contexts (representing 3157 surveyed fields). Contexts were constituted by surveyed and previous crops, rotations, soil-type, and production area. Source: Agreste – Enquête pratiques culturales 2011.

4 DISCUSSION

Although we have created typologies to group surveyed fields, the data loss was important for each analysis presented in this paper (Table 1). Moreover, those losses concern mostly the rarest TM and TS, as they are described on less fields. Therefore, ploughing is overrepresented in the study of TM repartition (on average 78% of surfaces for available data, compared to 64.9% on national average), and these results should only be used in a comparative way.

Data loss shall have consequences on qualitative factors describing the groups: it should be more difficult to describe a significant correlation between a TM or TS and a soil-type such as silt and loamy silt hydromorphic soil-type when it represents only 5 groups out of 194 for the PCA + AHC analysis.

4.1 NPT tillage correlated to fall-sown crops

Surveyed crop is the main factor correlated to chosen TM. NPT are mainly used before fall crops (cereal crops, rapeseed and pea). Their growth and potential yield are less impacted by soil structure and the quality of the seedbed, as these crops can compensate emergence losses (Labreuche et al., 2014). Moreover, intercrop period is shorter before these crops than before spring crops, which can have two main consequences: i) farmers have less time to make tillage operations (reducing working time during high activity peaks is one of the first quoted reason to change from plough to no-till in Europe – Soane et al., 2012); ii) seedbed quality may be not sufficient, as there may be not enough wetting and drying cycles for clods to be broken before sowing,
especially for soils with a tilled layer containing 20% clay or more (Labreuche et al., 2014).

Correlation between tillage methods and soil-type have also been established: reduced tillage (8–15 cm or 0–8 cm) is more used on calcareous clay soils, which seems coherent with Soane et al. (2012) and Labreuche et al. (2007) reviews. However, fall crops represent 57% of calcareous clay soils surfaces (52% of tillage crop surfaces in France), so there are still interactions between surveyed crop and soil-type.

Greater use of NIT and topsoiling in area 4 (South-West of France), mainly on supposed non-suitable crops (sunflower and corn) and non-suitable soils (hydromorphic silt & loamy silt) could be related with a high erosion risk existing in some parts of this area. This risk has been linked to ploughing in slope direction by Guiresse & Revel (1995), and may have caused NPT conversions faster in this area than in other parts of France (Chapelle-Barry, 2008).

4.2 Tillage strategies correlated to rotation-type

Ploughing & NPT strategies are the most used in the 2006–11 period (47% of tillage crop surfaces). The all-ploughing strategy represents 40% of surfaces, whereas the all-NPT strategy only 13% over the same period. Few farmers seems to give up ploughing completely, as using NPT is more or less easy, depending on conditions. Labreuche et al. (2014) stated that giving up ploughing implies farmers' adaptation and increase in technical skills.

On the one hand, cereal based rotations (spring crops or alternate fall and spring crops) are correlated to all-ploughing strategy. Labreuche et al. (2014) stated that cereal crops and corn are generally not suitable for NPT, as these crops produce large amount of residues that could have drawback for the following crop (delaying soil warming, preventing soil-seed contact). However, these negative effects of crop residues are depending on climate conditions (Soane et al., 2012), which could explain why we observe this correlation between ploughing and previous cereal crop only at a rotation scale (six-year period).

On the other hand, rotations with alternate cereal and dicotyledonous crops (significant for fall-sown crops) are correlated to all-NPT strategy. Short intercropping periods are numerous in these rotations and imply less time to plough and uncertainty about seedbed quality, as presented before (Soane et al., 2012; Labreuche et al., 2014).

Calcareous clay soils are correlated to all-NPT strategy. In the cluster where 25% of crop surfaces is under this strategy, areas 4 (South-West of France) and 7 (North-East) are also well represented. Calcareous clay soil-type represents respectively 40 and 45% of these areas (24% on national scale).

There is no negative correlation established between NPT and the soil-types supposed non-suitable for NOT use. This could be due to the dataset, with less surveyed fields on these soils due to a fewer capability for tillage crops. But the effects of the climatic conditions of the crop sequences could explain this result, as already stated by Soane et al., (2012), and NPT practices could actually enhance soil condition and capability to practice NPT on them (for instance, structure could be strengthened by higher OM content on soil surface).

4.3 Correlations between tillage and all other practices in the crop sequence

Correlations between NPT or ploughing and other cultural practices & crop performances are usually studied in field experiments, in specific pedoclimatic contexts. The metanalysis on no-till experiments made by Soane et al. (2012) indicated that the use of herbicides was necessary for the spread in Europe of this tillage method. Vacher and Colbach (2014) also indicated that ploughing can be used to control weed populations. This study confirmed a greater use of herbicides in NPT fields on average, and almost all this increase is dedicated to weed control.

Greater use of cover crops (CC) during intercrop period was an identified new practice that could reduce drawbacks/enhance advantages of NPT (Labreuche et al., 2007). CC could protect soils from erosion and protect soil structure (Bertuzzi et al., 2012), increase long term organic matter content (with effects on soil structure stability and on biological activity), and/or contribute to weed control (Charles et al., 2012). However, the proportion of surfaces with a CC during the intercrop period is nearly the same in NPT fields and in ploughed fields. Nevertheless, CC are more diversified and destructed 3 weeks later (which could allow the cover crop to produce more biomass) in NPT fields: hence, advantages for soil condition and soil capability provided by CC could be higher in NPT fields, as the positive effects of CC stated earlier strongly rely on cover crop species and agronomical performances (Charles et al., 2012; Bertuzzi et al., 2012). Soane et al., (2012) stated that increase in organic matter content in soil surface with no-till could enhance soil capability to produce with no-till.

NPT use led on average to small yield loss compare to ploughing (−5%), confirming results from Labreuche et al. (2014), even if important variability is stated because of crop and climatic conditions. As inputs are also higher on average in these fields, yield difference might be due to i) change of practices of farmers linked to change of TM; ii) farmers choice to use NPT tillage on soil with a smaller capability.

5 CONCLUSION

Tillage methods and tillage strategies distribution are mainly explained by surveyed crop and crop rotation: NPT are more used before fall crops in 2010–11 crop sequence and all-NPT strategy is correlated to fall-sown cereal and dicotyledonous crop rotation (as opposed to all-ploughing strategy for cereal-based rotations).

The use of reduced tillage in the 2010–11 crop sequence, and the use of the all-NPT strategy during the 2006–11 period are correlated to calcareous clay soils. However, no negative correlation was established between NPT and soil-types assumed not suitable for NPT: actually, the opposite result was found for hydromorphic silt & loamy-silt soil-type in the Souht-West of France. It could be caused by i) correlations between weather conditions and NPT drawbacks (drawbacks usually decrease with a drier climate); ii) other factors than soils taken into account by farmers to choose NPT (reducing working time, topography, etc.).

Compared to ploughed fields, NPT use is correlated to longer implantation of cover crops that may have positive effects on soil structure & weed control, and could cause a long-term increase of organic matter content of soils: this could enhance both soil condition and capability.

ACKNOWLEDGEMENTS

The "Agreste – Enquête Pratiques culturales 2011" survey data used in this paper has been provided by the CASD, thanks to the CAITTEC project led by ARVALIS – Institut du végétal.

REFERENCES

Agreste, 2014. Enquête Pratiques culturales 2011. Principaux résultats. *Agreste Les Dossiers* 21.

Bertuzzi P., Justes E., Le Bas C., Mary B., Souchère V. 2012. Effets des cultures intermédiaires sur l'érosion, les propriétés physiques du sol et le bilan carbone. *Réduire les fuites de nitrate au moyen de cultures intermédiaires. Conséquences sur les bilans d'eau et d'azote, autres services écosystémiques.* Paris: INRA.

Cavan N., Labreuche J., Wissocq A., Angevin F., Cousin I., submitted. Les cultures d'automne et les sols argilo-calcaires sont favorables à la mise en œuvre de techniques culturales sans labour. *Etude et Gestion des Sols.* Orléans: Association Française d'Etude des Sols (AFES).

Chapelle-Barry C., 2008. Dans le sillon du non-labour. *Agreste Primeur* 207.

Charles R., Montfort F., Sarthou J-P., 2012. Effets biotiques des cultures intermédiaires sur les adventices, la microflore et la faune. *Réduire les fuites de nitrate au moyen de cultures intermédiaires. Conséquences sur les bilans d'eau et d'azote, autres services écosystémiques.* Paris: INRA.

Colbach N., Vacher C. 2014. Travail du sol et gestion de la flore adventice. *Faut-il travailler le sol? Acquis et innovations pour une agriculture durable.* Versailles: Quae.

Guiresse M., Revel J.C., 1995. Erosion due to cultivation of calcareous clay soils on hillsides in south-ouest France. II. Effect of ploughing down the steepest slope. *Soil & Tillage Research* 35: 157–166.

Labreuche J., Lecomte V., Sauzet G., Leclech N., Longueval C., Martin C., Eschenbrenner G, Roger-Estrade J. 2014. Travail du sol et rendement des cultures: conditions et modalités de mise en œuvre pour les principales espèces de grande culture. *Faut-il travailler le sol? Acquis et innovations pour une agriculture durable.* Versailles: Quae.

Labreuche J., Viloingt T., Caboulet D., Daouze J.P., Duval R., Ganteil A., Jouy L., Quere L., Boizard H., Roger-Estrade J., 2007. La pratique des Techniques Culturales Sans Labour en France. *Evaluation des impacts environnementaux des Techniques Culturales Sans Labour (TCSL) en France.* Angers: ADEME.

Lê, S., Josse, J., Husson, F., 2008. FactoMineR: An R Package for Multivariate Analysis. *Journal of Statistical Software*, 25(1): 1–18.

Roger-Estrade J., Labreuche J., Boizard H., 2014. Importance du travail du sol: typologie des modes de mise en œuvre et effets sur le rendement des cultures. *Faut-il travailler le sol? Acquis et innovations pour une agriculture durable.* Versailles: Quae.

Soane, B.D., Ball, B.C., Arvidsson, J., Basch, G., Moreno, F., Roger-Estrade, J., 2012. No-till in northern, western and south-western Europe: A review of problems and opportunities for crop production and the environment. *Soil & Tillage Research* 118: 66–87.

Balancing decisions for urban brownfield regeneration: People, planet, profit and processes

L. Maring
Deltares, Utrecht, The Netherlands

F.L. Hooimeijer
Technical University Delft, Delft, The Netherlands

J. Norrman
Chalmers University of Technology, Göteborg, Sweden

ABSTRACT: In this paper, the soil security concept is applied on urban soils. In the SNOWMAN project Balance4P, the main focus was on the integration of subsurface aspects in urban brownfield regeneration. Balancing decisions for urban brownfield regeneration should take into account the three aspects of sustainability: people planet and profit; as well as the processes during the different phases in regeneration projects. A suggested framework gives concrete recommendations towards common practices of urban planners and subsurface engineers. This paper will guide along the Balance4P framework and the different aspects of "soil security".

1 INTRODUCTION

1.1 The importance of brownfield regeneration

Land take as a result of urbanization is one of the major soil threats in Europe. One of the key measures to prevent further urban sprawl and additional land take, is redevelopment of urban brownfields. Brownfields can be defined as sites affected by former uses and surrounding land; are derelict or underused; may have real or perceived contamination problems; are mainly in developed urban areas and; require intervention to bring them back to beneficial use (Ferber et al., 2006). The latter issue can be a bottleneck for redevelopment of brownfields instead of greenfields. A difficulty for brownfield redevelopments is that in urban projects the responsibilities, tools and knowledge of soil and subsurface engineering and urban planning and design are not integrated; they depend heavily on each other but work in different sectors. The urban designers usually deal with opportunities for socio-economic benefits while the soil and subsurface engineers deal with the technical challenges of the site.

1.2 Sustainable urban development

The global-wide trend of urbanization increases the importance of careful spatial planning in cities (OECD & CDRF, 2010). The sensibility of sustainable (re)development of urban areas is clear when considering climate change, population growth and increasing human demands for the living environment (Roberts & Sykes, 2000). In urban (re)development, sustainability has more recently gained awareness worldwide and sustainable development is quickly gaining in popularity (Lakkala & Vehmas, 2013). In literature, several reasons have been named for this sudden increase in popularity of sustainable development: bad practices have led to sub-optimal solutions and unsustainable situations; population growth and the depletion of natural resources call for a change in development practice; and sustainability is now a well-known marketing strategy (Kumar et al., 2012). This increasing trend in sustainable development can be seen in most aspects of society: food production, clothing, energy use, architecture, and in the spatial planning field. In order to prevent urban sprawl, decrease of property value and to increase the future livability of the city, the redevelopment of derelict and often contaminated land within the urban area is needed (Chakrapani & Hernandez, 2012).

1.3 Sustainable urban soil and subsurface management

In many cases, when talking about soil in urban areas, the scope is mainly on contamination and remediation. In the remediation sector, there is a broad on-going work to develop methods and tools that supports sustainable remediation. Remediation was earlier viewed as a sustainable action in itself, but today negative impacts of remediation are acknowledged. These negative impacts are for example transport emissions and fatality risks, health risks during remediation, consumption of energy and materials, as well as being costly. There is today an increasing demand for assessing remedial activities with regard to all three of the commonly mentioned sustainability dimensions: environment, economy and society. The International Standard Organization (ISO) currently works on a standard for sustainability evaluation of remedial actions (ISO, 2014) and there are several SuRF (Sustainable Remediation Forum) organizations worldwide that support this development. SuRF-UK suggested a general framework for assessing the sustainability of soil and groundwater remediation, broad enough to apply across different timescales, site sizes, and project types (Bardos et al., 2011). In accordance with Bardos et al. (2011), there are several attempts to incorporate sustainability in early phases of projects, as there is a general idea that the largest (sustainability) gains are achieved early in projects where they are still flexible.

Next to remediation, sustainable urban soil management is becoming more important in urban areas due to increasing pressure on space and climate change. The URBAN SMS project is advocating a broader view on soils in the city. Soil is a basis for space for housing, industrial and commercial purposes as well as infrastructure, recreational areas and food production. But soils in the urban environment also offer sustaining biological activity, diversity and productivity; regulating and partitioning water and solute flow; filtering, buffering, degrading, immobilizing and detoxifying of harmful substances originating from industrial and municipal by-products as well as from atmospheric depositions; storing and cycling nutrients and other elements within the earth's biosphere; production of renewable primary products; control of microclimate and mesoclimate, and providing support for socioeconomic structures and protecting archaeological treasures (Blumlein et al., 2012). Sustainable urban soil management means taking the broad range of soil functions into account.

1.4 *The Balance 4P project*

The background to the Balance 4P project is the idea that a better cooperation between urban developers and subsurface specialists in early phases of the redevelopment process can accelerate brownfield redevelopment and potentially identify more sustainable redevelopment strategies. A major aim was to develop a holistic approach that supports sustainable urban renewal through the redevelopment of contaminated land and underused sites (Norrman et al., 2015a; 2015b; 2016). Balance 4P was developed in close cooperation with three case studies in the three countries involved: Sweden, Belgium and The Netherlands. In the project were, next to the end-users, soil and subsurface engineers and urban planners involved.

2 THE BALANCE 4P HOLISTIC FRAMEWORK

2.1 *What is the Balance 4P framework?*

The Balance 4P project focuses on the planning process (initiative and plan phases). The project focuses on the common practices of urban planners and subsurface engineers to be able to find the opportunities to improve their processes and the collaboration between these groups. An analysis was done of the planning systems in the participating countries: Sweden, Belgium and The Netherlands on the local, regional and national scales (Hooimeijer & Tummers, 2017). Law, policy, regulation and institutions determine the planning conditions that have to be taken into account during urban brownfield redevelopment projects. The urban brownfield development process can be divided in two distinct processes: the planning process, consisting of two major phases: initiative and planning phase; and the implementation process, consisting of the realization and maintenance phase. Each of these phases are associated with different stakeholders, timelines, working processes, information needs and instruments, but all of them typically have a divergent part, where information is gathered and convergent part, where directions are set and decisions are made (Figure 1).

Next to the analysis of the planning systems, an inventory of policies and regulations for soil and

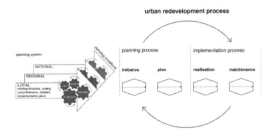

Figure 1. The holistic approach is operating within planning conditions that are the result of all levels in the planning system (local, regional, national) and their respective laws and regulations, policy and institutions.

subsurface was performed to find out what aspects of soil are mandatory to take into account during the planning process and which are optional. In the three studied countries, best opportunities for subsurface integration in current planning systems were found in: heritage, environment, nature and water (Figure 2).

The proposed Balance 4P framework (Figure 3) consist of four major steps, for which the potential supporting tools and instruments, the "technical" output and the intended learning output are given. The steps are described in the following section.

2.2 The Balance 4P framework steps elaborated

The first step is a broad stakeholder analysis in which not only the stakeholders are named, but also their interest in the issue: those interests that will be affected by the decision to be taken. The resources that the stakeholder possesses that can be used in the decision making (knowledge, information, leverage, money) are also inventoried, next to the point if the stakeholder can mobilize these resources quickly or slowly. This is important when looking at the dynamics of the decision making. If a decision needs to be taken quickly, but the resource (e.g. knowledge) can only be delivered slowly, this resource is of less importance than previously thought. Also their position on the issue is of importance, to determine how to approach them. People can be neutral, strongly or slightly negative or positive. Strongly negative stakeholders take a lot of energy and will in many cases not be convinced. However, a way to handle this opposition (reduce negative impact) is necessary in the strategy for decision making. For the slightly negative stakeholder, a convincing argument could be enough to become neutral or (slightly) positive. The slightly and strongly positive stakeholders can be activated and sustained to support the issue (Maring, 2013). Note that during the urban brownfield development the stakeholder analysis should be updated or redone, because the stakeholders will change during time and phase at hand.

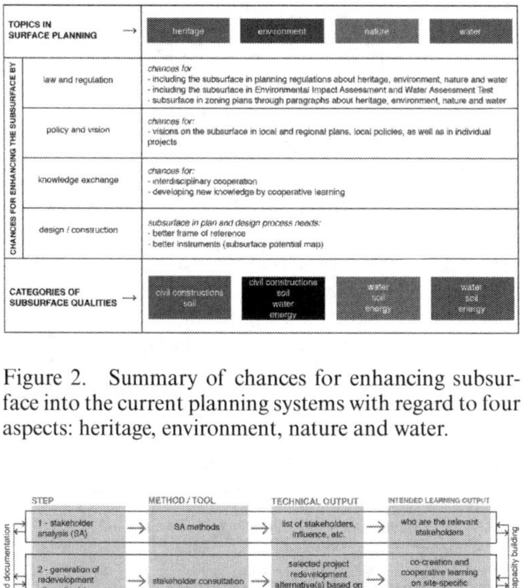

The second step is the regeneration of redevelopment alternatives. This phase is about gathering information and investigation of what demands, objectives and boundary conditions exist for the urban brownfield regeneration. This is done for the "aboveground" matters. How many houses, offices, amenities and other functions are needed, what objectives exists in the area in terms of meeting societal needs (food, drinking water, energy production, shelter, infrastructure) and overcoming societal challenges (climate change mitigation and adaptation, increasing demands on non-renewable natural resources, environmental justice). This is also done for soil and subsurface qualities. These qualities consists of challenges (problems or boundary conditions) for the aboveground developments and of opportunities (soil functions and services) that can be employed within the urban brownfield regeneration. Examples of the challenges are contamination, low carrying capacity of the soil, existing subsurface constructions or infrastructures that might hinder new constructions, or unexploded ordnance. Examples of opportunities are: possibility to use aquifer thermal energy storage (ATES), water storing capacity in the upper layer of the soil to mitigate peak discharges, production capacity for green areas and parks, and favorable building conditions. To investigate this connection between the aboveground and subsurface, the stakeholders should be involved. For this the, System Exploration Environment and Subsurface (SEES) was developed, Figure 4 (Hooimeijer & Maring, 2013). Note that the subsurface topics are divided in themes that recognizable and appeal to urban developers: civil structures, energy, water and soil (green). Output of this step is project redevelopment alternatives based

Figure 2. Summary of chances for enhancing subsurface into the current planning systems with regard to four aspects: heritage, environment, nature and water.

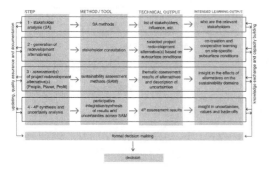

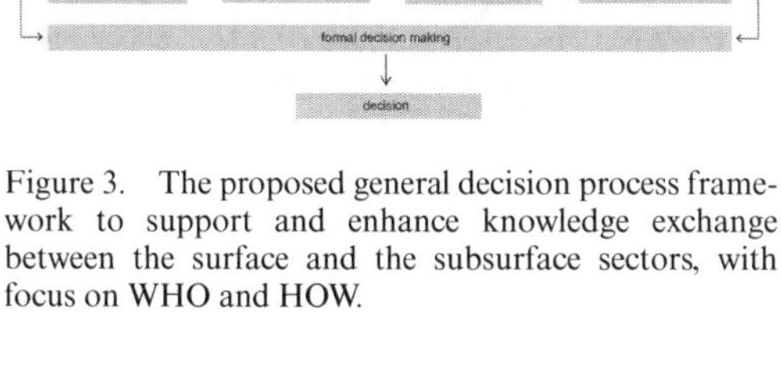

Figure 3. The proposed general decision process framework to support and enhance knowledge exchange between the surface and the subsurface sectors, with focus on WHO and HOW.

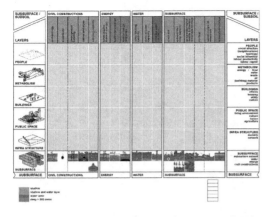

Figure 4. System Exploration Environment and Subsurface SEES, with attention to different knowledge fields and experts on aboveground and subsurface topics. SEES is available at: https://publicwiki.deltares.nl/display/SEES/HOME+EnglishTIMBRE.

on subsurface conditions. Outcome of this step is the co-creation, the networking and the cooperative learning on site-specific surface and subsurface conditions and ambitions.

Step three consists of the assessment of project redevelopment alternatives. The assessment can be done balancing the aspects people, planet and profit. For assessing project alternatives, many different tools were created in different countries, during different projects (Norrman et al, 2015; 2016). Of importance is that it is clear to the users, how the assessment is done, so its outcomes will be accepted by stakeholders involved. The output of this step is the thematic assessment results of alternatives and description of uncertainties. The outcome is the insight in effects of alternatives on the three sustainability domains: people, planet and profit.

The fourth and final step is the synthesis and the uncertainty and gap analysis. This step is a transparent analysis of overlaps, missing data and uncertainties, and critical aspects based on the assessments of the project alternatives. It is recommended to involve relevant stakeholders. During or after this step, it can be decided that there is not enough information for decision making, or that certain tradeoffs determine that a specific stakeholder should be involved more strongly or in a different way. The output of this step is the assessment result and the outcome is insight into uncertainties, values and trade-offs.

These steps does not form a linear, but a circular and iterative process. During the steps, it is crucial to keep updating, pay attention to the quality of information gathered and the activities being performed and to document all information of importance, especially these which were or will be involved in decision-making. Next to that, knowledge exchange and capacity building is a continuous activity when performing these steps. The outcome of the framework is decision support and based on this, the formal decision-making process can start, and a final decision on an urban redevelopment alternative be delivered.

2.3 *The Balance 4P project: Conclusions and challenges*

The Balance 4P project concluded the following. Next to "people, planet and profit", the fourth P: the process is crucial for the success of sustainable projects. This includes a process leader who stays on top of the objectives and boundary conditions that are set for the project. This includes being specific that all parties involved should take subsurface into account when performing their tasks. Otherwise this will end up on the end of the list as "extra work", or "complicating" or "difficult".

Knowledge exchange is key to success in sustainable urban brownfield redevelopment projects. For the knowledge exchange it is not only of importance of what knowledge is delivered, also how and when. People should be capable of understanding the information delivered and be able to use it for their specific task. Therefore it is also important not to overload people involved with all information in the beginning of the project, but dose the information supply: the right information to the right person in the right time, and explain when needed. Knowledge exchange is a continuous activity that needs planning.

There is a multitude of available tools (Kok, 2014; Maring et al., 2013) and the choice which tool fits best depends on data availability, project phase, ability, legal framing but also personal taste. The most important aspect when choosing a tool is that stakeholders involved accept the tool and its outputs.

A good design matters for sustainable solutions. It is worth to spend the perceived extra effort in the beginning of the project for a better, more resilient and robust design. This avoids unexpected events (Lackin et al., 2014) and pays off in the maintenance phase.

Remediation is just one subsurface aspect in urban planning and design. When involving the soil expert, it is in many cases the person who knows about contamination. Important aspect is attention at the beginning of the project to which other soil and subsurface experts to involve: e.g. engineers with knowledge of the hydrogeological, geotechnical and ecological aspects, soil energy, and civil engineers.

One final conclusion of the Balance 4P project is that the soil, subsurface and planning sector can benefit from each other's practices.

There are also challenges identified to implement the holistic framework in practice. There is a lack of regulatory and policy support for inclusion of subsurface in the planning phases. When there is no rule or policy that asks for the effort to take soil and subsurface into account, it solely depends on the project/process manager if (s)he values this topic as important.

The quality of the information transfer is not secured during the redevelopment process when the actors and/or regulatory frameworks change. In many cases huge information loss is the result.

There is limited interest of stakeholders and planners for subsurface inclusion and sustainability assessments in early planning due to complexity. Examples of what taking soil and subsurface into account during the planning process contributes to in terms of avoided costs and damages and better plans, can take away the hesitations.

In most cases there is limited planning project budget and unbalanced risk cost distribution between developers and planners. This also hinders the application of the holistic framework. Other ways of tendering and contracting (e.g. Design Build Maintain contracts) can help to overcome this obstacle.

3 SOIL SECURITY APPLIED TO THE URBAN BROWNFIELD REDEVELOPMENT PROCESS

The concept of soil security is divided in five "C's" (McBratney, 2014) 1) Capability, 2) Condition, 3) Capital, 4) Connectivity, and 5) Codification. In this section, these C's are compared with the Balance 4P steps.

1. Capability: What functions does the soil have? This is one of the main objectives of applying the SEES instrument together with soil and subsurface engineers and urban planners. Beforehand the experts are asked to investigate the information on the site that will be (re)developed and the potential functions of the soil are taken along. The planners also prepare the aboveground demands (which land use functions, societal challenges and needs are of importance for the project).
2. Condition: what is requested in terms of land use functions, societal challenges and needs, and what can the soil deliver? In this step, the connection is between the aboveground and the soil-subsurface system. Where there is a mismatch between the two, solutions should be found in different design or technical solutions.
3. Capital: using land for hard uses and infrastructures will provide most income, but in (re)-development areas, other values will be of importance to reassure the success of a project. Next to profit, also the people and planet aspects of sustainability are important. In the Balance 4P project, the ecosystem services value of soil is particularly taken into account in the planning process. More examples of this value can help to promote the approaches of planning with soil and subsurface.
4. Connectivity: This dimension is crucial for realizing urban planning which considers soil and subsurface functions. When the land owner, project leader or important stakeholders do not know or care about soil and subsurface, soil and subsurface information will not be used in the planning process. Again: good examples can help to improve connectivity dimension.
5. Codification: policy, regulation, legislation, governance tools, but also initiatives and examples are all drivers that can help to support taking soil and subsurface into account in urban redevelopment projects. However, as can be seen in the Balance 4P conclusions in section 2.3, there are few existing today. Urban planning with soil and subsurface is an emerging field of expertise.

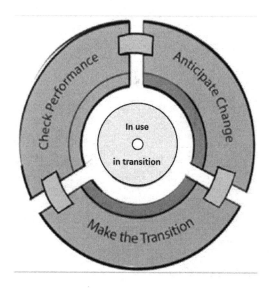

Figure 5. Circular land management: the administrative land management cycle (outer cycle) addresses land use transitions in the land use cycle (inner cycle) (van Gaans & Ellen, 2014).

4 CONCLUSION

The soil security concept and the Balance4P framework seem to aim at the same objective: take the full range of functions of soil and subsurface into account when using land and soil.

Good land, soil and subsurface management is when we are using all five dimensions of soil security. This has a similarity with circular land management as developed and promoted in the Circuse (Preuß, Thomas & Verbücheln, 2013), HOMBRE (van Gaans, 2014) and Balance 4P (Norrman et al., 2015) projects. Circular land management consists of continuous management of land. In circular land management the performance of land is monitored and action is taken, ideally before mismatches occur. In this way, land is prevented to be at risk of becoming a brownfield again.

Soil security is mainly applied on agricultural land, but it seems to be perfectly applicable to urban land as well. Soil and subsurface functions in urban land are still quite unexplored. The first concepts such as the Balance 4P framework and soil security concept are not yet picked up by the urban planners: there is a gap to bridge. However, there seem to be opportunities to make big steps in sustainability in urban areas adopting these concepts. Research projects and practical examples will contribute to this.

REFERENCES

Bardos, R.P., Laurent M.M. Bakker, Hans L.A. Slenders, and C. Paul Nathanail Sustainability and Remediation (2011). Chapter 20 in Swartjes F.A. (ed.) (2011) Dealing with Contaminated Sites, DOI 10.1007/978-90-481-9757-6_20, C_Springer. Science + Business Media B.V.

Coffin, S.L., (2003). Closing the Brownfield Blumlein P., H.J. Kircholtes, M. Schweiker, G. Wolf, B. Schug, I. Wieshofer, Sigbert Huber, M. Parolin, F. Villa, A. Zelioli, M. Biasioli, P. Medved, Tomaz Vernik, Borut Vrščaj, Grzegorz Siebielec, Josef Kozák, I. Galuskova, E. Fulajtar, Jaroslava Sobocká, S. Jaensch, 2012 Soil in the City. Urban Soil Management Strategy. ISBN: 978-3-943246-07-0, City of Stuttgart – Department for Environmental Protection, Germany.

Chakrapani, C., & Hernandez, T. (2012, 06). Brownfield redevelopment and the triple bottom line approach. Retrieved 29–04–2014, from www.mah.gov.on.ca: http://www.mah.gov.on.ca/AssetFactory.aspx?did=9658.

Ferber, U., Grimski, D., Millar, K., Nathanail, P., 2006. Sustainable Brownfield Regeneration: CABERNET Network Report. ISBN 0-9547474-5-, University of Nottingham, UK.

Hooimeijer, F.L. & Maring, L., 2013. Ontwerpen met de Ondergrond. in: Stedenbouw & Ruimtelijke Ordening 2013/6.

Hooimeijer, F.L. and Tummers, L. (2017) Harmonizing subsur-face management in spatial planning in the Netherlands, Sweden and Flanders. ICE Urban Design & Planning journal. ISO Guide 73:2009, 3.2.1.1 https://www.iso.org/obp/ui/#iso:std:iso:guide:73:ed-1:v1:en.

Kok, J.J., 2014. A Guide Through the Forest of Sustainable Urban Redevelopment Instruments. Internship report VU Amsterdam, Deltares, The Netherlands.

Kumar, V., Rahman, Z., Kazmi, A., & Goyal, P. (2012). Evolution of sustainability as marketing strategy: Beginning of new era. Procedia- Social and Behavioral Sciences, 37, 482–489.

Lackin, J., Stuurman, R., Bleeker, R., 2014. Onverwachte gebeurtenissen in de bodem. TCB R23(2014), The Netherlands.

Lakkala, H., & Vehmas, J. (2013, 10 10). Trends and Future of Sustainable Development – Proceedings of the Conference 'Trends and Future of Sustainable Development' 9–10 June 2011, Tampere, Finland. Turku, Varsinas-Suomi, Finland.

Maring, L., Blauw, M., van den Bergh, R., Ellen, GJ., van Oostrom, N., van Gaans, P., Limasset, E., Ferber, U., Menger, P., Smit, M., Wendler, K., Neonato, F., Irminski, W., Nathanail, P., Barros Garcia, R., 2013. HOMBRE HOMBRE D 3.1: Decision support framework for the successful regeneration of brownfields.

McBratney, A., Field, D.J., Koch, A., 2014. The dimensions of soil security. Geoderma 213, pp. 203–213.

Norrman, J., Volchko, Y., Maring, L., Hooimeijer, F., Broekx, S., Garção, R., Kain, J.-H., Ivarsson, M. Touchant, K., Beames A., 2015a. Balance 4P: Balancing decisions for urban brownfield regeneration – technical report. Report 2015:11. Department of Civil and Environmental Engineering, Chalmers University of Technology. Gothenburg, Sweden.

Norrman, J., Maring, L., Hooimeijer, F., Broekx, S., Garção, R., Volchko, Y., Kain, J.-H., Ivarsson, M. Touchant, K., Beames A., 2015b. Balance 4P: Balancing decisions for urban brownfield regeneration – case studies. Report 2015:12. Department of Civil and Environmental Engineering, Chalmers University of Technology. Gothenburg, Sweden.

Norrman, J., Volchko, Y., Hooimeijer, F., Maring, L., Kain, J-H., Bardos, P., Broekx, S., Beames, A., Rosén, L., 2016. Integration of the subsurface and the surface sectors for a more holistic approach for sustainable redevelopment of urban brownfields. Science of the Total Environment, 563–564, pp. 879–889.

OECD, & CDRF. (2010). Trends in Urbanisation and Urban Policies in OECD Countries: What Lessons for China? doi: 10.1787/9789264092259-en: OECD publishing.

Preuß, T. & Verbücheln, M., (Eds.), 2013. Towards Circular Flow Land Use Management. The CircUse Compendium, Berlin.

Roberts, P. & Sykes, H., 2000. Urban Regeneration: a handbook. Wilts: Cromwell Press.

van Gaans, P. & Ellen, GJ., 2014. HOMBRE D 2.3: Successful Brownfield Regeneration.

Extension of irrigation in semi-arid regions: What challenges for soil security? Perspectives from a regional-scale project in Navarre (Spain)

Rodrigo Antón, Iñigo Virto, Jon González, Iker Hernández, Alberto Enrique & Paloma Bescansa
Soil Science Lab, Department of Sciences of the Natural Environment, ETSIA, Public University of Navarre (UPNA), Pamplona, Spain

Nerea Arias
Soil Science Lab, Department of Sciences of the Natural Environment, ETSIA, Public University of Navarre (UPNA), Pamplona, Spain
Department of Innovation, Section of Sustainable Agrosystems, Navarre Institute of Agrifood Infrastructures and Technology, Avda. Serapio Huici, Villava, Spain

Luis Orcaray
Department of Innovation, Section of Sustainable Agrosystems, Navarre Institute of Agrifood Infrastructures and Technology, Avda. Serapio Huici, Villava, Spain

Raquel Campillo
GAP Recursos, Ltd. Puente la Reina, Navarre, Spain

ABSTRACT: The conversion from dryland to irrigation in semi-arid land is a widespread strategy to grant agricultural profitability and food security. The trade-offs of this transformation for soil security remain unclear. The project LIFE REGADIOX, based on the establishment of a regional-scale network of representative plots in three irrigation districts in Navarre (NE Spain), allowed for a rational evaluation of soil security by using the soils capability to establish fair comparisons in terms of soil condition, capital and connectivity in irrigated vs rainfed plots. The results showed a clear influence of irrigation in soil condition and capital, arising from greater SOC storage. Differences in other soil indicators were uneven, and related to the natural limitations of the sites studied (soil and climate), and to the time under irrigation. The translation of these changes into the soil capital showed that irrigation adoption can altered the soils capacity to provide key ecosystem services beyond biomass production, as enhanced greenhouse emissions and moderated changes in soil erodibility were recorded. Connectivity was studied from differences in the major driver for farmers' decisions on soils (gross gains), and compared to SOC gains. Results showed that economic (income) and environmental (SOC gain) drivers did not always match. All in all, our results indicate that soil security can be affected in opposite directions when irrigation is implemented in semi-arid dryland. The dimension of these changes depends on the natural soil characteristics, and the management conditions of the agrosystems. Optimizing soil management under irrigation seems essential for ensuring a positive evolution of soil security in this and other semi-arid regions.

1 INTRODUCTION

The challenge of transforming agriculture to feed a growing population without harming natural resources such as soil, requires a transition towards agricultural systems able to supply enough goods without losing sustainability. This implies a more efficient use of inputs, more yields stability and greater resilience to risks, crises and climate variability in the long term (FAO, 2017). The conversion from dryland to irrigation in semi-arid land is a widespread strategy in this sense, and is often seen as an efficient tool towards food security (Darko et al., 2016), as it reduces yield loss caused by drought and water stress, and increases flexibility in crop planting dates and crop types (Salmon et al., 2015).

Globally, 2.3–4.0 Mkm² or 15–26% of the global croplands are equipped for irrigation (Erb et al., 2016). This accounts for 33–40% of global food production (Salmon et al., 2015), including 44% of total cereal production, 30% of the global wheat fields (0.7 Mkm²), 20% of the maize fields (0.3 Mkm²) and half of the global citrus, sugar cane and cotton crops (Portmann et al., 2010).

Although the expansion of irrigation has slowed down in the last years, due to higher competition for water resources (FAO, 2017), Southern Europe has seen its irrigated surface growing in the last decades. In 2013, the total irrigable area in the EU-27 was 18.644 Mha, which represents an increase of 13.4% compared to 2003. Spain was the second country where the share of irrigable area increased the most, from 16.8% to 31.1% (Eurostat, 2013). In absolute terms, Spain and Italy had the largest irrigable areas (6.7 and 4.0 Mha, respectively). In addition to surface extension, the average irrigation requirements estimated by Wriedt et al. (2008) are highest in the Mediterranean area, reflecting the general climatic characteristics of this region, as well as soil attributes and crops demands.

The region of Navarre (NE Spain) is an example of the expansion of irrigation in semi-arid Mediterranean land. For a total agricultural surface of 341,835 ha, more than 22,300 ha of rainfed land was converted to irrigation in the last years (2000–2015), which added to the previous existing irrigated area make a total of 108,221 ha of irrigated land in 2016 (Gobierno de Navarra, 2017a). In 2009, irrigated land production in Navarre already accounted for 35% of agricultural gross domestic product (GDP).

In general, irrigation adoption implies major changes in agrosystems, as it means changes not only in crop yields and profitability, but also in fertilization, rotations and soil management, and in the use of water and energy. For instance, irrigation affects the energy and radiation balance at the surface (Salmon et al., 2015). As such, irrigation has been identified within the ten important land management activities that may impact the Earth system profoundly (Erb et al., 2016).

Many of these changes can affect soil security, defined as being concerned with the maintenance and improvement of soil to produce food, fiber and freshwater, contribute to energy and climate sustainability, and maintain the biodiversity and the overall protection of the ecosystem (McBratney et al., 2012). Irrigation can indeed alter soil functions related to global societal challenges: food, water and energy security, climate change abatement, biodiversity protection, and the maintenance of ecosystem services (McBratney et al., 2014).

Because of these sometime conflicting consequences, the role of irrigation in rural development can have contradictory views, arising from the heterogeneities between the preferences of stakeholder groups regarding water resources management, agricultural practices, and irrigation challenges. Although for some, irrigation can be the cornerstone of future adaptation strategies in face of climate and socio-economic changes in Europe (Dunford et al., 2015), integrating the assessment of the trade-offs between different services (e.g. energy and food security, Hurford & Harou, 2014)), and the provision of ecosystem services other than productivity is needed for a better understanding of these issues (e.g. Aspe et al., 2016; Ricart & Clarimont, 2016). For instance, Dominati et al. (2017) showed that including the modifications of soil condition following the adoption of irrigation was necessary for an accurate assessment of ecosystem services and their net present value (NPV) at a regional scale.

In this context, the project LIFE REGADIOX (http://life-regadiox.es/en/) aimed to design, test, and spread the impact that an improved model of sustainable management of irrigated agriculture can have in climate change in the region of Navarre (NE Spain). The project was designed to assess the potential consequences of different management strategies in irrigated agriculture in relation to the main global environmental challenges defined by McBratney (2014) at a regional scale, including greenhouse gases (GHG) balances in irrigated soils, and the optimization of energy, water and other inputs consumption. Based in the establishment of a regional-scale network of representative plots, it also allowed for a rational evaluation of soil security in at least four out of its five dimensions (*capability, condition, capital* and *connectivity*), as it provided a multi-disciplinary framework including the study of inherent and manageable soil properties (Dominati et al., 2010), monitoring of environmental outcomes of different management strategies in irrigated land, and socio-economic analyses.

In this work we evaluated soil security from the perspective described above in three irrigation districts in Navarre, in which different agricultural managements exist, including rainfed and irrigated crops, using an approach based in the assessment of soil *condition, capital* and *connectivity*, after defining the local reference state of soil *capability* at each location.

2 MATERIALS AND METHODS

2.1 *Irrigation districts and fields*

The study was conducted on a series of agricultural plots in three irrigation districts (Miranda, Funes and Valtierra), selected considering the particularities of climate and soils in the region. In relation to climate, the three districts were representative of the aridity gradient existing in the region from North to South.

At each of these districts, representative agricultural plots were selected to stablish a network that would be used for evaluating soil properties, GHG emissions and production and monetary

Table 1. Sites characterization.

Site	Miranda de Arga	Funes	Valtierra
Coordinates	42°28′59″N 1°49′36″W	42°18′51″N 1°48′10″W	42°11′43″N 1°38′03″W
Mean Annual Precipitation (mm)	437	390	379
Mean Annual Temperature (°C)	13.9	14.0	13.9
Potential evapotranspiration (Thornthwaite) (mm)	738	740	740
Reference soil (S.S.S., 2014)	*Typic Calcixerept*	*Xeric Haplocalcid*	*Xeric Haplocalcid*
Managements included in the study	Dryland Annual irrigated Irrigated fodder	Dryland Annual irrigated	Dryland Annual irrigated Organic dryland
Years since conversion to irrigation	6	13	20

parameters. This was done by identifying plots with a known historic management, including rainfed cereal cropping since at least 50 years, and the most widespread cropping systems in irrigated land for 6, 13 and 20 consecutive years (since irrigation was implemented) in Miranda, Funes and Valtierra, respectively. In particular, the plots selected included rainfed wheat and irrigated corn at the three sites, plus organic rainfed barley in Valtierra and fodder (alfalfa) in Miranda de Arga.

From these fields, soil parameters were evaluated using an *ad hoc* sampling design (see 2.2), and annual management data (including fertilization doses and timing, tillage and all other inputs and operations, as well as yields) were collected from farmers managing the fields (see 2.4).

A summary of this soil and climate characterization is shown in Table 1.

2.2 *Soil capability and sampling design*

Following McBratney et al. (2014), soil capability was studied at each site for an adequate assessment of the other dimensions of soil security, and the *local reference state* was defined by considering both the genoform and phenoform of soils (*sensu* Droogers & Bouma, 1997). Genetic information of the soils at each site was collected from available public databases on soil, including soil cartography (Soil map of Navarre 1:25,000), geology (Geologic map of Navarre 1:25,000) and orto-photographs 1:5,000 available at the regional public database (https://idena.navarra.es/Portal/Inicio), and from soil pits sampled for this study. From these data, pedological criteria were used to ensure that the selected plots included soil from the most representative taxonomic unit at each site (*Xeric Haplocalcid* in Valtierra and Funes, and *Typic Calcixerept* in Miranda, S.S.S., 2014).

Then, the reference state was set for soils under permanent non-irrigated cereal cropping, because dryland agriculture has been practiced for decades (or centuries), inducing non reversible changes in soils due to erosion and soil organic matter losses, and resulting in a phenotypical alteration of the taxonomically defined genoform.

Within these plots, a sampling strategy was developed to ensure that only soils with similar capabilities were compared at each site. That for, since many of the selected fields included soils from more than one cartographic unit, the sampling zone in each plot was defined by adjusting it to only the surface corresponding to the characteristic soil type by image processing using SIG software (QSIG). In all cases, the process was completed with a field visit to verify the final result and with extra analysis when necessary. Hence, the sampling network for soils included only those areas within the reference soil at each plot (Figure 1).

Finally, to ensure a representative sampling, a protocol was especially designed from the one proposed by Stolvoboy et al. (2007), as described by de Soto et al. (2017). This includes a sampling scheme based on a randomized template that is adjusted to the sampling area of the studied plot. Its use allowed for establishing a geo-referenced grid including three representative areas (n = 3) per sampling zone and plot. At each area, one composite soil sample from 25 sub-samples and other intact soil core for bulk density (BD) were collected in the tilled layer at 0–15 cm and 15–30 cm (Figure 1).

2.3 *Soil analyses*

A minimum dataset of soil physical, biological and organic matter-related properties was defined for assessing soil condition based on previous studies

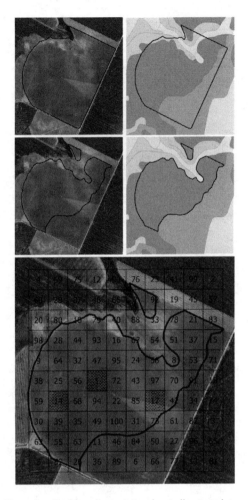

Figure 1. Surface adjustment and sampling template.

developed in the region for the selection of soil quality indicators (Fernández-Ugalde et al., 2009; Imaz et al., 2010). These included BD and the soil water-holding capacity (SWHC), microbial biomass C (MBC), and an index of soil functional biodiversity (see below), total (SOC) and particulate organic C (POM-C).

Pressure plate extractors (Soil Moisture Equipment Corp., Santa Barbara, CA) were used for SWHC determinations, as described by Dirksen (1999). The SWHC was calculated from the difference in soil moisture at field capacity (−33 kPa) and permanent wilting point (−1500 kPa). Volumetric values for SWR were calculated from the gravimetric measures using BD.

Because of the presence of carbonates, wet oxidation (Walkley-Black) was used to analyse total oxidizable C (Tiessen and Moir, 1993), from which we calculated total and particulate organic C (POM-C), considered that contained in the fraction >53 μm (Cambardella and Elliot, 1992).

Microbial biomass C was determined by the fumigation-extraction method of Vance et al. (1987). The functional diversity of the soil microbial population was studied through the analysis of the community-level physiological profiles (CLPLs). The C source utilization patterns observed using a Biolog Ecoplate™ microplating system (Biolog, Hayward, CA, USA) were used to determine functional diversity of soil microorganisms. These plates comprise 31 C substrates that are major ecologically relevant compounds. The number of substrates used by the soil microbial community (NSU) was used as indicator.

Finally, soil erodibility was determined by calculating the K factor of the Revised Universal Soil Loss Equation (RUSLE) model (Renard et al., 1991). It represents the susceptibility of soil to erosion, and it depends on the inherent soil properties (texture, organic matter, structure and permeability).

2.4 Data collection for dry biomass production and GHG emissions

Farmers responsible for the management of the selected plots participated in a survey designed to collect all management data relative to one average season. An exhaustive database including yields and all production factors including tillage and other operations, fertilization and irrigation doses and other inputs was completed. From yield data in the survey, the annual production of dry biomass was calculated based on relative humidity of each product (14% for maize and 12% for wheat, barley and alfalfa).

The GHG emission rates were estimated with a software developed in the framework of a regional project for C footprint calculations in agriculture (EURENERS3, 2017). According to the requirements of PAS 2050 (2011), the calculation includes both direct (generated directly during the production process), and indirect emissions (those occurring in a different location but linked to the production process, such as the generation of electric power, production and transport of inputs or treatment of plastic waste). The calculator was fed from the farmers' survey database. Results were provided in eq Mg CO_2 per ha corresponding to emissions associated to soil respiration, N fertilization, tillage, crops residues burning, energy for irrigation and those associated to other inputs.

2.5 Cost-efficiency

Financial information on the gross gains of each studied plot was obtained from the farmers' survey, or from market prices (Gobierno de Navarra, 2017b) when not provided. This allowed for a comparison in income (gross gain) of the different management systems studied at each location, in relation to the reference state (dryland cereal).

Previous studies on the introduction of irrigation (Dominati & Mackay, 2015) showed that the final impact on soil connectivity needs to address not only the monetary benefits of improved production, but also the evaluation of such benefits in relation to the changes observed in soil condition. Because SOC gains were the most evident difference observed in soil condition in irrigated plots compared to the reference state (see below 3.1.), we chose to represent also the ratio between the changes observed in gross gain and in SOC. Therefore, a cost-efficiency indicator was obtained by calculating the gross economic gain of each irrigated system in comparison to the reference system (dryland wheat), per unit (Mg) of SOC sequestered, on a *per ha* annual basis:

$$\text{Gross gain}\left(\frac{\text{€}}{ha}\right) \text{per} \frac{Mg}{ha} \text{of C sequestered} = \frac{\Delta \text{gross gain (irrigated - reference)}}{\Delta SOC \text{ (irrigated - reference)}}$$

3 RESULTS AND DISCUSSION

3.1 *Condition*

Irrigation is known to potentially affect soil health (e.g. Adejumobi et al., 2016), although this affection is site- and management-dependent. Among other consequences, introducing irrigation implies changes in the soil water regime, alters nutrients cycling and significantly modifies the C cycle (Apesteguía et al., 2015).

In this study, the assessment of soil condition was done by comparing soil indicators for irrigated management (and organic dryland in Valtierra) with the reference state (soil under dryland cropping) at each district (Table 2). The conversion from dryland to irrigation had different consequences in these soil parameters. In general, an increase in SOC, considered a 'universal' indicator for soil condition (Stockmann et al., 2013; McBratney et al., 2014) was observed in all soil types. This trend was also observed, to a lesser extent, in the dryland plot with organic amendments.

Similar results have been observed in other regions when irrigation is adopted, and are seen as a consequence of greater productivity and crops C incorporation into the soil (e.g. Gillabel et al., 2007; Denef et al., 2008). In this sense, in the review of Zhou et al., 2016 on 179 published studies,

Table 2. Soil condition (mean ± std. dev. n = 3).

Site		Physical condition			Biological condition				Organic matter	
		Bulk density (kg/dm³)		SWHC* (mm)	Microbial biomass-C (µg C g soil⁻¹)		Number of substrates used		POM-C* (Mg/ha)	SOC* (Mg/ha)
	Prof. (cm)	0–15	15–30	0–30	0–15	15–30	0–15	15–30	0–30	0–30
Miranda de Arga	Reference	1.7 ± 0.1	1.7 ± 0.1	533 ± 46 a	379 ± 110	326 ± 124 a	26.3 ± 4.2	28.0 ± 2.6b	5.1 ± 1.8 a	43.9 ± 4.0a
	Annual irrigated	1.7 ± 0.1	1.5 ± 0.0	659 ± 58 b	290 ± 68	204 ± 50 b	24.7 ± 2.1	22.3 ± 1.5a	6.4 ± 0.4 a	56.4 ± 2.8b
	Irrigated fodder	1.8 ± 0.0	1.6 ± 0.0	626 ± 21 b	237 ± 85	222 ± 15 b	24.7 ± 1.5	25.3 ± 2.1b	6.2 ± 1.5 a	63.1 ± 7.0c
Funes	Reference	1.6 ± 0.0	1.7 ± 0.1	588 ± 35 a	238 ± 24	252 ± 102	9.33 ± 4.2 a	8.0 ± 3.5a	4.5 ± 0.7 a	35.4 ± 0.7a
	Annual irrigated	1.6 ± 0.2	1.6± 0.1	605 ± 60 a	243 ± 29	226 ± 5	20.0 ± 2.6 b	20.0 ± 1.7b	7.7 ± 0.5 b	46.3 ± 0.7b
Valtierra	Reference	1.5 ± 0.3	1.7 ± 0.1	539 ± 7 a	338 ± 55	152 ± 22 a	22.0 ± 6.6 a	17.7 ± 3.1a	4.4 ± 1.7 a	57.3 ± 4.2a
	Organic dryland	1.4 ± 0.1	1.5 ± 0.1	470 ± 87 ab	488 ± 159	448 ± 217 b	14.0 ± 3.0 a	17.7 ± 3.1a	11.4 ± 0.7 b	74.5 ± 9.5b
	Annual irrigated	1.6 ± 0.1	1.7 ± 0.1	351 ± 85 b	223 ± 76	160 ± 10 a	31.3 ± 0.6 b	27.0 ± 1.0b	18.4 ± 4.1 c	99.6 ± 9.7c

*SWHC: Soil water-holding capacity; POM-C: C in the particulate organic matter; SOC: Total soil organic C. Different letters in the same column and site indicate significant differences (p < 0.05).

irrigation irrigation increased SOC by 1.27% on average. This was related to an increment in aboveground and belowground net primary production of 25.5% and 31.4%, respectively. The observed differences can be compared with previous studies on the effect of land-use changes on SOC: Trost et al. (2013) compiled 5 studies and reported average net gains in SOC of +17.6% in irrigated land vs. dryland in semi-arid conditions, with a high variability. Our data of +41.3% on average (maximum and minimum of +73.9% and +28.3% in annual irrigated in Valtierra and Miranda de Arga, respectively) fall above those values. These values corresponded to 1.49 ± 0.14, 2.64 ± 0.10 and 2.07 ± 0.10 Mg SOC ha^{-1} year^{-1} in Valtierra, Miranda and Funes, respectively. For comparison, average SOC gains following NT adoption in Spain were set at 0.72 ± 0.16 Mg SOC ha^{-1} per year by González-Sanchez et al. (2012), suggesting that irrigation adoption is a more efficient strategy in this sense.

The site-dependency of these changes were however seen when studying POM-C, which can be considered an earlier and more sensitive indicator of changes in soil organic fraction than total soil C (Imaz et al., 2010). POM-C increased with irrigation, but not in all cases: no differences were observed in Miranda de Arga, the district with the shortest time after conversion, but also the one with the most reduced water deficit (Table 1). This site also showed losses in MBC at 15–30 cm, while Zhou et al. (2016) reported average gains of 42.2%. As for POM-C, changes in soil biodiversity were also uneven among sites, with the Miranda site showing none or negative effects, and Funes and Valtierra positive effects in annual crops. Seasonal and site-dependent differences in the soil microbial community gains with irrigation were also observed by Calderón et al. (2016), when comparing two sites with irrigation doses in the US.

Finally, in relation to the soils' physical condition, we did not observe significant differences in BD at any site, which could be explained because all the soils studied were under a similar soil till management. SWHC results were however uneven, showing better condition in irrigated land only in Miranda.

Overall, this results show that soil condition was affected by the adoption of irrigation, and that this affection was site- and management-dependent.

3.2 *Capital*

The soil's natural capital is determined by the compositional state of the soil system, which affects the functions provided by the soil for the whole ecosystem (McBratney et al., 2014). In this work, we focused on three major ecosystem services of agricultural soils, considering the characteristics of the region: two regulating services (climate regulation and erosion control), and the most significant provisioning service in agrosystems (biomass production).

In relation to biomass, as expected, the conversion to irrigation implied a remarkable increase in yields and dry biomass production. In contrast, the adoption of organic production without irrigation in Valtierra resulted in reduced yields.

Climate regulation can be evaluated by comparing the net gain in SOC resulting from the change in management (CO_2 abatement) with GHG emissions generated by each agrosystem (in tons of equivalent CO_2, Table 3). The results indicated that irrigation increased SOC, but also GHG emissions. Previous studies (Erb et al., 2016) have already pointed out that irrigation can induce greater GHG emissions because it affects soil moisture, temperature and N availability, which are all drivers for the production and evolution of GHG emissions from soils. Sanz-Cobena et al. (2017) explored the different strategies for GHG emissions mitigation in Mediterranean agriculture. It is clear from their study that irrigated systems create favorable soil conditions for N_2O production. In this study, however, the differences observed are the result of different inputs and soil management system, since GHG emissions were estimated from these data, and not actual soil measurements. Emissions associated to N fertilization accounted for most of the emissions in our study plots (data not shown). Because emission factors in these systems fluctuate greatly according to water management and the type and amount of fertilizer used, water management practices and the adjustment of N fertilization seem the most promising practices for mitigation, as shown by Karimi et al. (2012), Ptle et al. (2016) and Zornoza et al. (2016).

Soil erosion is a key indicator in the assessment of ecosystem services provision in irrigated land (Herzig et al., 2016). The *K* coefficient in the RUSLE equation is usually reduced with SOC gains. However, as its calculation also depends on soil texture and structure, some interference with these factors made that the observed gains in SOC were translated in reduced erodibility only in Valtierra. It has to be added that irrigation changes also at least the *R* (rainfall erosivity) and *C* (crop cover) factors in the equation, so that the final effect on erosion should also consider these changes.

All in all, these results showed that irrigation adoption can alter also the soil capital and its capacity to provide key ecosystem services beyond biomass production, as observed by Dominati et al. (2016). The net effect will depend on multiple factors, including soil management strategies. The optimization of the soil capital will therefore depend on the optimization of these parameters.

Table 3. Soil capital & connectivity.

Site		GHG* emissions (Mg eqCO$_2$ ha^{-1} y^{-1})	Erodibility (K) (Kg ha h ha^{-1} MJ^{-1} mm^{-1})	Biomass Production (kg dry matter ha^{-1})	SOC* gains Total (Mg ha^{-1})	SOC* gains Average (Mg ha^{-1} y^{-1})	Gross gain (€ ha^{-1})	Δ gross gain/ΔSOC
Miranda de Arga	Reference (dryland)	1.24	44.9 ± 0.8[a]	3872	–	–	545.4	–
	Annual irrigated	4.50	49.9 ± 0.5[b]	13149.4	12.4 ± 2.8	2.07 ± 0.4	1911.5	661.0
	Irrigated fodder	0.83	57.9 ± 2.5[c]	11440	19.2 ± 4.7	3.20 ± 0.8	1132.9	183.6
Funes	Reference (dryland)	0.67	63.3 ± 0.1[b]	3918	–	–	376.2	–
	Annual irrigated	7.44	61.8 ± 0.5[a]	11591	10.9 ± 4.3	0.84 ± 0.3	1569.0	1422.5
Valtierra	Reference (dryland)	0.67	52.5 ± 0.5[c]	4840	–	–	755.0	–
	Organic dryland	0.77	49.5 ± 2.3[b]	290	17.2 ± 6.01	0.86 ± 0.3	508.3	–
	Annual irrigated	5.36	44.3 ± 3.9[a]	11610	42.3 ± 6.12	2.12 ± 0.3	1434.5	321.3

3.3 Connectivity

Connectivity was evaluated considering the changes in gross monetary gains of each agrosystem, as it represents the major driver for decisions related to soil management in the context of this study. While the transformation to organic management resulted in net losses, very likely due to the reduction in yields, irrigation represented a net increment of income in all the studied sites (Table 3). When this gain was compared to SOC gains, it was observed that there was not a direct correlation between management strategies resulting in SOC gains and economic gains (for instance, irrigated fodder vs. annual irrigated in Miranda de Arga). This means that higher absolute and relative income increase (which represent the best option for farmers) did not correspond to higher SOC gains. Similarly, in Valtierra, organic farming, which represented the lowest increase in GHG emissions in comparison to reference state, displayed negative gross gains. These two cases are an example of how economic and environmental drivers of farmers' decisions can induce different results in soil security. Soil codification through regulation can be important in this sense if win-win strategies, implying gains in income and ecosystem services, are to be implemented. Policies encouraging the adoption of managements increasing soil capital and condition should consider the need for financial compensation if this is the major driver for decision makers in relation to soil management.

Another result is that values for the same soil use (annual irrigated) were different for different sites and time under irrigation, because SOC gains were site- and time-dependent (Table 3) for similar biomass productions. The evaluation of soil connectivity should therefore account for these conditions.

4 CONCLUSIONS

This work showed that the major changes in management occurring when dryland agricultural soils are converted to irrigation can induce significant changes in soil condition, capital and connectivity, and therefore in the soils functionality (McBratney et al., 2014) and their ability to provide adequate ecosystem services.

The most evident change in soil condition were net gains in SOC. Given that SOC may play a role in mitigating GHG emissions and in delivering major soil ecosystem services and functions (Ogle & Paustian, 2005), policymakers are increasingly focusing their attention on measures for SOC conservation. However, this study shows that other changes in the capacity of soils to provide ecosystem services need to be accounted for when a holistic evaluation of the consequences of irrigation adoption are to be assessed at a regional scale, especially from the perspective of optimizing soil security in all its dimensions. On the other hand, other management strategies, such as organic agriculture, can have similar consequences in soil condition and capital, but worse performances in terms of biomass production and profitability. Soil codification, as represented by regulations affecting the decisions of soil managers, seems an interesting tool to ensure soil security in this and other semi-arid regions where irrigating is being expanded.

REFERENCES

Adejumobi, M.A., Awe, G.O., Abegunrin, T.P., Oyetunji, O.M., Kareem, T.S. 2016. Effect of irrigation on soil health: a case study of the Ikere irrigation project in Oyo State, southwest Nigeria. *Environtal Monitoring Assessment* 188: 696.

Asociación TEDER 2011. EURONERS3, version 3. http://teder.org/proyectos/ohuella-de-carbono-de-producto-agroalimentario-eureners-3/

Aspe, C., Gilles, A., Jacque, M. 2016. Irrigation canals as tools for climate change adaptation and fish biodiversity management in Southern France. *Regional Environmental Change* 16:1975–1984.

Apesteguía, M., Virto, I., Orcaray, L., Enrique, A., Bescansa, P. 2015. Effect of the conversion to irrigation of semiarid mediterranean dryland agrosecoystems on soil carbon dynamics and soil aggregation. *Arid Land Research and Management* 29(4):399–414.

Cambardella, C.A. & Elliott, E.T. 1992. Particulate soil organic matter changes across a grassland cultivation sequence. *Soil Science Society of America Journal* 56(3):777–783.

Darko, R.O., Yuan, S., Hong, L., Liu, J., Yan, H. 2016. Irrigation, a productive tool for food security – a review. Acta Agriculturae Scandinavica, Section B— *Soil & Plant Science*, 66(3): 191–206.

de Soto, I.S., Virto, I., Barre, P., Fernandez-Ugalde, O., Anton, R., Martinez, I., Chaduteau, C., Enrique, A., Bescansa, P. 2017. A model for field-based evidences of the impact of irrigation on carbonates in the tilled layer of semi-arid Mediterranean soils. *Geoderma* 297:48–60.

Denef, K., Stewart, C.E., Brenner, J., Paustian, K. 2008. Does long-term center-pivot irrigation increase soil carbon stocks in semi-arid agro-ecosystems? *Geoderma* 145: 121–129.

Dirksen, C. 1999. *Soil physics measurements*. Catena Verlag. GeoEcology Paperback.

Dominati, E., Patterson, M., Mackay, A. 2010. A framework for classifying and quantifying the natural capital and ecosystem services of soils. *Ecological Economics* 69 (9): 1858–1868.

Dominati, E., Mackay, A. 2015. Impact of on-farm built infrastructure investments on the provision of ecosystem services: irrigation for dairy systems in New Zealand. In: Currie L.D., Burkitt, L.L. (eds) *Moving farm system to improved attenuation.* http://flrc.massey.ac.nz/publications.html. Occasional Report No. 28. Fertilizer and Lime Research Centre. 10–12 February 2015. Massey University, Palmerston North.

Dominati, E., Mackay, A., Bouma, J., Green, S. 2016. An Ecosystems Approach to Quantify Soil Performance for Multiple Outcomes: The Future of Land Evaluation? *Soil & Water Management & Conservation* 80:438–449.

Dominati, E., Mackay, A., Rendel, J. 2017. Understanding soils' contribution to ecosystem services provision to inform farm system analysis. In Field, D.J., Morgan, C.L.S., McBratney, A.B. (Eds) Global Soil Security. Coll. Progress in Soil Science. Springer International Publishing. Switzerland. Pp. 207–217.

Dunford, R.W., Smith, A.C., Harrison, P.A., Hanganu, D. 2015. Ecosystem service provision in a changing Europe: adapting to the impacts of combined climate and socio-economic change. *Landscape Ecology* 30:443–461.

Erb, K.H., Luyssaert, S., Meyfroidt, P., Pongratz, J., Don, A., Kloster, S., Kuemmerle, T., Fetzel, T., Fuchs, R., Herold, M., *et al.*, 2016. Land management: data availability and process understanding for global change studies. *Global Change Biology* 23: 512–533.

EURENERS 3, 2017. Huella de carbono de producto agroalimentario EURENERS 3. Available at http://teder.org/proyectos/ohuella-de-carbono-de-producto-agroalimentario-eureners-3/ [In Spanish]. (Accessed on May 2017).

Eurostat, 2016. Agri-environmental indicator – irrigation. Available at http://ec.europa.eu/eurostat/statistics-explained/index.php/Agri-environmental_indicator_-_irrigation. (Accessed on May 2017).

FAO. 2017. The future of food and agriculture – Trends and challenges. Rome

Fernandez-Ugalde, O., Virto, I., Bescansa, P., Imaz, M.J., Enrique, A., Karlen, D.L. 2009. No-tillage improvement of soil physical quality in calcareous, degradation-prone, semiarid soils. *Soil & Tillage Research* 106 (1): 29–35

Gillabel, J., Denef, K., Brenner, J., Merckx, R., Paustian, K. 2007. Carbon sequestration and soil aggregation in center-pivot irrigated and dryland cultivated farming systems. *Soil Science Society of America Journal* 71: 1020–1028.

Gobierno de Navarra, 2017a. Superficies agrícolas. Available at https://www.navarra.es/home_es/Temas/Ambito+rural/Vida+rural/Observatorio+agrario/Agricola/Informacion+estadistica/superficies+agricolas.htm. [In Spanish] (Accessed on May 2017).

Gobierno de Navarra, 2017b. Precios y mercados. Available at https://www.navarra.es/home_es/Temas/Ambito+rural/Vida+rural/Observatorio+agrario/Agricola/Informacion+estadistica/Temas+economicos.htm [In Spanish] (Accessed on May 2017).

Gonzalez-Sanchez, E.J., Ordonez-Fernandez, R., Carbonell-Bojollo, R., Veroz-Gonzalez, O., Gil-Ribes, J.A. 2012. Meta-analysis on atmospheric carbon capture in Spain through the use of conservation agriculture. *Soil & Tillage Research* 122:52–60.

Herzig, A., Dymond, J., Ausseil, A-G. 2016. Exploring limits and trade-offs of irrigation and agricultural intensification in the Ruamahanga catchment, New Zealand. *New Zealand Journal of Agricultural Research*, 59(3):216–234

Hurford, A.P., Harou, J.J. 2014. Balancing ecosystem services with energy and food security – Assessing trade-offs from reservoir operation and irrigation investments in Kenya's Tana Basin. *Hydrology Earth System. Sciences* 18: 3259–3277.

Imaz, M.J., Virto, I., Bescansa, P., Enrique, A., Fernandez-Ugalde, O., Karlen, D.L. 2010. Soil quality indicator response to tillage and residue management on semi-arid Mediterranean cropland. *Soil & Tillage Research* 107(1): 17–25.

Karimi, P., Qureshi, A.S., Bahramloo, R., Molden, D. 2012. Reducing carbon emissions through improved irrigation and groundwater management: A case study from Iran. *Agricultural Water Management* 108:52–60.

McBratney, A.B., Minasny, B., Wheeler, I., Malone, B.P. 2012. Frameworks for digital soil assessment. In: Minasny, B., Malone, B.P., McBratney, A.B. (Eds.), *Digital Soil Assessment and Beyond* 9–14. Taylor & Francis Group, London.

McBratney, A.B., Field, D.J., Koch, A. 2014. The dimensions of soil security. Geoderma 213: 203–213.

Ogle, S.M. & Paustian, K 2005. Soil organic carbon as an indicator of environmental quality at the national scale: Inventory monitoring methods and policy relevance. *Canadian Journal of Soil Science* 85(4):531–540.

PAS 2050, Specification for the assessment of the life cycle greenhouse gas emissions of goods and services, BSI report (2011).

Patle, G.T., Singh, D.K., Sarangi, A., Khanna, M. 2016. Managing CO2 emission from groundwater pumping for irrigating major crops in trans indo-gangetic plains of India. *Climatic Change* 136(2):265–279.

Portmann, F.T., Siebert, S., Döll, P. 2010, MIRCA2000- Global monthly irrigated and rainfed crop areas around the year 2000: A new high-resolution data set for agricultural and hydrological modeling, Global Biogeochemical Cycles, 24, GB1011.

Quantum GIS Development Team, 2015. Quantum GIS Geographic Information System, version 2.12. Open Source Geospatial Foundation Project. http://qgis.osgeo.org.

Renard, K.G., Foster, G.R. Weesies, G.A., Porter, J.P. 1991. RUSLE: Revised universal soil loss equation. *Journal of Soil and Water Conservation* 46(1): 30–33.

Ricart, S., Clarimont, S. 2016. Modelling the links between irrigation, ecosystem services and rural development in pursuit of social legitimacy: Results from a territorial analysis of the Neste System (Hautes-Pyr_en_ees, France) *Journal of Rural Studies* 43:1–12.

Salmon, J.M., Friedl, M.M., Frolking, S., Wisser, D., Douglas, E.M. 2015. Global rain-fed, irrigated, and paddy croplands: A new high resolution map derived from remote sensing, crop inventories and climate data. *International Journal of Applied Earth Observation and Geoinformation* 38: 321–334.

Sanz-Cobena, A., Lassaletta, L., Aguilerac, E., del Prado, A., Garnier, J., Gillen, G., Iglesias, A., Sáncheza, B., Guardia, G., Abalos, D., Plaza-Bonilla, D., *et al.* 2017. Strategies for greenhouse gas emissions mitigation in Mediterranean agriculture: A review. *Agriculture, Ecosystems and Environment* 238:5–24.

Stockmann, U., Adams, M.A., Crawford, J.W., Field, D.J., Henakaarchchi, N., Jenkins, M., Minasny, B., McBratney, A.B., de Courcelles, V.D., Singh, K., *et al.* 2013. The knowns, known unknowns and unknowns of sequestration of soil organic carbon. *Agriculture Ecosystems & Environment* 164:80–99.

Stolbovoy, V., Montanarella, L., Filippi, N., Jones, A., Gallego, J., & Grassi, G., 2007. Soil sampling protocol to certify the changes of organic carbon stock in mineral soil of the European Union. Version 2. EUR 21576 EN/2. 56 pp. Office for Official Publications of the European Communities, Luxembourg.

Soil Survey Staff (SSS), 2014. Keys to Soil Taxonomy, 12th ed. USDA-Natural Resources Conservation Service, Washington, DC, USA

Tiessen, H., Moir, J.O. 1993. Total and organic carbon. In M.R. Carter (eds), *Soil Sampling and Methods of Analysis*: 187–200. CRC Press LLC, Boca Raton, FL, USA

Trost, B., Prochnow, A., Drastig, K., Meyer-Aurich, A., Ellmer, F., Baumecker, M. 2013 Irrigation, soil organic carbon and N2O emissions. A review. *Agronomy for Sustainable Development* 33(4): 733–749.

Vance, E.D., Brookes, P.C., Jenkinson, D.S. 1987. An extraction method for measuring soil microbial biomass-C. *Soil Biology & Biochemistry* 19(6):703–707.

Wriedt, G., Van der Velde, M., Aloe, A., Bouraoui, F. 2008. JRC Scientific and Technical Reports. Water Requirements for Irrigation in the European Union. Office for Official Publications of the European Communities, Luxembourg.

Zhou, X., Zhou, L., Nie, Y., Fu, Y., Du, Z., Shao, J., Zheng, Z., Wang, X. 2016. Similar responses of soil carbon storage to drought and irrigation in terrestrial ecosystems but with contrasting mechanisms: A meta-analysis. *Agriculture, Ecosystems and Environment* 228: 70–81.

Zornoza, R., Rosales, R.M., Acosta, J.A., de la Rosa, J.M., Arcenegui, V., Faz, A., Pérez-Pastor, A., 2016. Efficient irrigation management can contribute to reduce soil CO2 emissions in agriculture. *Geoderma* 263:70–77.

Connectivity and raising soil awareness

Soil security to connectivity: The what, so what and now what

Damien Field
Sydney Institute of Agriculture, The University of Sydney, NSW, Australia

ABSTRACT: With the development of the Soil Security concept, and the articulation of the dimensions Connectivity, it is timely to rethink approaches to its teaching. There is a continued need to develop those who will have expertise in soil science, i.e. 'know' soil, but with the diversification of soil science beyond its historical focus on agriculture, there is a need to understand what this means for those who only want to 'know of' soil. This requires soil science to connect with other cognate disciplines and adopt a multidisciplinary approach to understand how non-soil scientists are connected to soil. To be inclusive soil science also needs to recognise those who see soil as part of their narrative, the artist, poet, citizen scientist, or advocates of social media. This group can be described as 'aware of' soil and develop connections through concepts of its care. This will require a rethinking the soil science curriculum and how it is delivered. This chapter explores the needs of the different learners of soil and uses some examples of delivered curriculum to illustrate the creation of a modern learning environment.

1 INTRODUCTION

The concept of soil security describes five dimensions; capability, condition, capital, connectivity and codification, which are used to monitor, analyse and discuss the challenge of food and water security, contribute to maintaining biodiversity and support human health (McBratney et al., 2014). Of these connectivity is probably the least developed dimension. This dimension compliments the need to place a value on the soil, i.e. its capital, and is concerned with understanding how society connects to soil (McBratney & Field, 2015).

A link has been made between soil security and the seven functions that soil provide (Bouma & McBratney, 2013). It is through these functions that the connection of society be mapped to determine 'what' the people's relationships with soil are and through a process of reflection (Rolfe et al., 2001) decide on the expert knowledge that is need to problem solve (Bouma et al., 2011). For example, farmers and graziers have the strongest connection to soil and are concerned with biomass production, soil function 1, but they may also adopt various management strategies to conserve the natural resource on farm contributing to soil functions number 3, biodiversity and 6, soil carbon. This well-established connection is maintained by those who work with their soil and 'know of' its capabilities and are concerned with maintaining its condition and value.

Farmers and land managers are supported by 'knowledge brokers' who provide advice on soil issues and ensure the extension of the latest soil knowledge (Dominati et al., 2010; Bouma et al., 2011). What is crucial is these brokers of soil knowledge have the expertise to 'know' soil combined with the social intelligence to understand why soil is relevant to those who know of soil, i.e. this is identified as the 'so what' in any problem-solving situation. It is crucial that the development of future soil science curriculums incorporate 'real-world problems' so that graduates have had the opportunity apply soil knowledge and skills and be able to integrate this knowledge with other considerations, such as; the economic feasibility, availability of resources, and abilities and needs of the end-user (Janzen et al., 2011; Bouma & McBratney, 2013; Hartemink et al., 2014; Field et al., 2017).

Finally, reconnecting society to soil is an immediate challenge and is predicated on those who 'care' about soil (McEwan et al., 2017). This concept of caring for soil may be founded on the notions of beauty and utility (Yaalon, 1996; McBratney et al., 2017). As a utility, society will increasingly appreciate and want to know the soil that provides good quality clean food, and into the future, a source of their pharmaceuticals (Sojka, 2002; Tugel et al., 2005; Robinson et al., 2012). Soil also provides spaces for people to recreate and the value here can be described as a contributor to human health and aesthetics (Keesstra et al., 2016). This group of people could be described as those who are 'aware for soil', whose valuing of soil does not rely on any expertise but believe in its need to exist and its protection for future generations, i.e. notions of care and a bequest value (Field & Sanderson, 2017).

The 'now what' (Rolfe et al., 2001) is concerned with the future and the actions that need to be taken now to facilitate this. There is always the continued need for improved soil data acquisition and its integration with other data sources to secure soil as a producer of food and water, and maintain biodiversity (Kidd et al., 2015). This data needs to be relevant for policy development and enable a value to be placed on soil, as those using soil are concerned with earning and compliance (Koch et al., 2013). Other institutions are also considering the value of soil, e.g. the banking sector through natural capital, as well as contributing to their social licence (Morgan et al., 2017). To do this they also need good quality and relevant soil data.

Changes to the soil education have been flagged (Bouma & McBratney, 2013) to ensure people who have good soil knowledge (Field et al., 2017) and can engage with the what, so what and now what when making connections with soil. Soil science education was framed around agriculture in the early 19th century (Brevik & Hartemink, 2010) but this connection has diversified and has moved beyond to include ecosystem services (Robinson et al., 2012), policy (Keesstra et al., 2016), and its role in aesthetics (Toland & Wessolek, 2010; Feller et al., 2015) This will require focusing on developing an agile and diverse curriculum (Field et al., 2017) that can facilitate those who will need to 'know' soil, 'know of' soil and/or who are 'aware of' soil.

2 'KNOW', 'KNOW OF' AND/OR BE 'AWARE OF', SOIL

There is a need to have soil experts that have a deep knowledge in the discipline. This is not only need for the continuing maturing of the discipline since its inception around 150 years ago, but also ensures that those who need good soil science knowledge can get salient advice (Bouma et al., 2011; Field et al., 2017). As illustrated in Table 1 this group who know soil are identified as discipline experts and are concerned with primarily the study of soil as an object and classically are categorised into the sub-disciplines such as pedology, soil chemistry & physics, or classification.

The curriculum is usually dominated by developing soil science knowledge and associated technical skills. Recent attempts of working with stakeholders to develop a soil science core-body knowledge resulted in 23 discipline-centred items ranging from the theoretical to the applied, with only 5 of these that could be classified as having a multidisciplinary focus (Field et al., 2017). Historically this knowledge was realised through soil's interaction with agriculture (Table 2).

Table 1. Basic characteristics of learner engagement who need to 'know', 'know of' or be 'aware of' soil.

Learning engagement	Characteristics
'Know'	Identified as the fundamental knowledge that is the discipline of Soil Science: soil as an object of study or science of soil materials; soil defined as a collection of properties; or soil represented by attributes & processes; soil genesis; communicated as a taxonomy
'Know of'	Identified as a component of knowledge that is useful in cognitive disciplines, e.g. agronomy, ecosystem services, natural capital; soil as a utility and/or provides a function, represented as a collection of indicators; with some awareness of soil attributes
'Aware of'	Identified by an amateur knowledge of soil; valuing soil through concepts of care; represented by its aesthetics; soil as a character in a larger narrative.

Table 2. The progression of soil science from amateurs and its development as a discipline to the recent demands for its multidisciplinary engagement and future inclusion of amateurs, (modified Koppi et al., 2010).

Aware of amateur	Know disciplinary expertise	Know of interdisciplinary expertise	Aware of amateur
The study of soil for utilitarian and survival purposes (over 150 years ago)	The study of soil in its own right; development of theories, concepts, methods and analytical procedures; a strong focus on agriculture.	Linking soil science with other disciplines and engaging with other scientists, politicians and stakeholders to provide information and solutions to complex environmental issues and problems	Members of the public who engage with soil science out of personal interest and enthusiasm for the discipline and its practices.

Application of this knowledge is often demonstrated by how it usefulness in understand threats to soil, such as; erosion acidification, contamination, loss of organic matter, compaction, and other physical degradation (McBratney et al., 2014), which usually form the basis of most soil science texts. Recent developments in; soil mapping with advance in pedometrics and digital soil mapping (Malone et al., 2013), derivation of remotely sourced soil data along with non-invasive techniques (Viscarra-Rossel et al., 2006; along with increasing use of soil data in modelling require a good grounding in soil knowledge (Bouma et al., 2011) and will contribute to its future disciplinary focused development. It is expected that such a curriculum would enable graduates to have the fundamental knowledge and skills that enable them to be recognised by soil science accreditation schemes (SSSA, 2007; CPSS, 2017).

Hartemink et al. (2014) describes a range of approaches to teaching soil science using some 15 examples from University undergraduate teaching from across the globe. While this diverse range of examples illustrate cultural and personal differences in their approaches to teaching it also illustrated that there is an increasing trend of teaching outside of the discipline (Havlin et al., 2010), and required students to engage in multidisciplinary thinking. This ability for graduates with soil knowledge to be able to make connections between soil with other disciplines, and the needs of society, was also demonstrated when Field et al. (2011) proposed a set of learning outcomes specific to soil science. Work reported by Kelly et al. (2006), Jarvis et al. (2012) and Field et al. (2013) have reported the need to soil science graduates to have good discipline knowledge but also the skills to communicate and engage with stakeholders outside of the discipline (Table 2). This engagement is focused on demonstrating the relevance of knowledge and should go beyond the just knowing the science but also be able to incorporate the societal demands and align with their values and ethics (Nowotny, et al., 2002). As illustrated in Table 1 the engagement of soil is more than likely focuses on its utility and trying to assess its role in the functions that it can provide. This curriculum will often focus on problem-solving framework to explore the knowledge of the discipline. Of equal importance is the opportunity to explore the economic and social dimensions that are affected by soil. As described earlier this might involve connecting soil's role as an ecosystem service (Robinson et al., 2014), its contribution to value through estimations of natural capital (Dominati et al., 2017; McBratney et al., 2017), or its presence in forming policy (Napier, 1998). In this learning environment, it is expected that the graduate's social intelligence will be developed so they can make better informed advice to those who need good soil science knowledge that is relevant to their demands. As put by Bouma et al. (2011), it may not be about the right or wrong but about considering the decision through a lens of better or worse. The knowledge brokers educated through this curriculum will develop the ability to facilitate collaboration between researchers, the education community and end-users of soil knowledge (Stockmann et al., 2013). With the future development of soil specific policy and regulation there may also be the need to further develop accreditation for soil knowledge brokers that is not only recognised nationally but internationally facilitating transfer of soil experts globally.

In parallel Feller et al. (2015) have been investigating the role of soil in society as an aesthetic, while McEwan et al. (2017) reports on children's engagement with soil as part of understanding nature (White, 2008; Debenham, 2015). The value of soil is 'measured' through concepts of care and can be described as forming part of a narrative (Table 1). Those who are aware of soil can be identified as amateurs and according to Mims (1999) science has a long history of amateurs who do science, demonstrated by those who survey birds, tag butterflies or report astral phenomena. While often used as a pejorative term the French root for the word amateur, amour, is love and this often characterises this group. Their engagement with the discipline is driven by personal interest pure enthusiasm (Table 2), without formal training necessarily in discipline itself. Unlike some other sciences, such as geology and astronomy, the amateur does not feature often in societies of soil (Koppi et al., 2010). With the growing popularity of citizen science this may change in the not to distance future (Shelley et al., 2013). Hartemink et al. (2013) reported on curricula developed by van Rees where students can explore soil through art whereby science and arts students work together to explore and reflect on soil as a medium not just a science. Likewise, Field et al. (2017) found that some end-users want a learning environment where the needs of the industry determine what soil science knowledge needs to be learnt, rather than learning about soil remote from the challenges being faced. The development of community gardens and kitchen gardens are also examples of emerging soil curricula that raise people's awareness of soil (Kitchen Garden Foundation, 2015). Soil science should be mature enough to embrace these approaches where the entry point of study is not from the discipline of soil. In doing so this provides an opportunity for others to become aware of soil and according to Alex McBratney 'those who know care, and those who care lobby' (McEwan et al., 2017).

3 SOME EXEMPLARS OF RETHINKING THE CURRICULUM

Most traditional approaches to a curriculum at Universities will have a strong focus on lectures and practical classes in its early stages. In doing so the teachers decide and control the work to be completed and students are rarely given the opportunity to determine that learning environment (Field et al., 2017). The inclusion of field-school often in later years can provide an opportunity for students to be more active in their learning by giving them opportunities to decide the topics and methods to pursue (Figure 1), in effect a shift towards a negotiation between student and teacher (Grabineger & Dunlop, 1995). As reported by Field et al. (2013) this shift is identified by graduates and employers as an effective learning environment.

Soil profile description and subsequent classification is one area that can be challenging when trying to engage students. Developing these skills is not only important for those who need to know soil, i.e. demonstrate discipline expertise, but provides information that will support those who are giving advice to people who want to know of soil. Since 1961 a long-standing tradition in the USA has been the encouraging students to participate in the National Soil Judging competition (SSSA, 2017). Aide (1989) recognised that soil judging enable students to demonstrate their critical thinking. The inclusion of a T.E.A.M approach to soil judging (Cooper, 1991) further increased the learning experience where students now had to substantiate their judgement to the team, developing their communication skills and learn from shared experiences. In 2014, the first of the global soil judging competitions was held in Jeju, Korea (Cattle et al., 2014). According to Cattle (2014), soil judging is a valuable educational tool and seems to be a great way to engage a new generation of soil scientists. This enthusiasm may be party explained by the competitive nature of the activity balanced by the development of a team spirit, which is often difficult to achieve when a more traditional approach to soil description and classification are implemented.

As illustrated in Figure 1 taking the opportunity to develop 'research-based' practical experiences can benefit student to conceptualise soil science beyond just a set of irrefutable facts and anabolise observed during their practical experiences (Ala Samarapungavan & Bodner, 2006). The key to this approach is where groups of students are included in the decision process of sampling, choice of interrelated experiments, and reporting to solve a problem that they themselves have negotiated with the lecturer. Field et al. (2007) reported the implementation of this approach where students undertook a series of interrelated experiments on a soil profile they had sampled in the field and subsequently collated and interpreted the resulting data for publication in the Australian Journal of Soil Research (AJSR) now Soil Research. Essential to this experience was the groups keeping keep their own laboratory book describing their experiments, record how they overcame difficulties they encountered, and, through consultation with each other and the literature, try to explain any observed anomalies. Assessment was guided by the criteria used as a reviewer for AJSR. Laboratory books reflected a critical approach to the methods, with student results suggesting that they were questioning the dependability of the methods and their outcomes rather than their own technical abilities. Student surveys indicated that the practical experience improved student's ability to identify fundamental relationships between concepts and experimental observations, and an increased confidence in identifying and addressing problems encountered during experimentation (Field et al. 2007).

Representatives from 15 countries reported on their experiences in teaching soil science (Hartemink et al., 2014) and a common experience is the focus on initial attention in creating students interest in soil and once caught following this with courses that deepen their knowledge. As described here earlier this could tap into their competitive spirit or their interest in negotiating the learning experience. Incorporating a field-based component and considering the role of soil in global

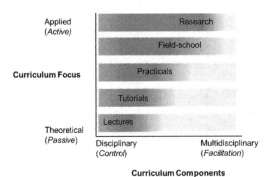

Figure 1. The teaching and learning approaches commonly utilised by soil science and its relationship with student engagement, curriculum focus from passive to active, and teacher participation, curriculum components from control to facilitation.

issues including climate change, food production, or environmental risks, along with the opportunity to use advanced technologies will deepen the students experience and learning (Hartemink et al., 2014). The opportunity to work on real-world problems so students can experience soil science and its incorporation in the jobs of the future, a discipline that where the demand for its expertise is increasing (Havlin et al., 2010; NERC, 2012).

The adoption of problem-based learning is proven to be an effective way of deepening students learning (Hartemink et al., 2014) and when using authentic real-world problems will engage students who need to 'know', as well as, students who need to 'know of' soil science (Field et al., 2011). Table 3 illustrates a generic module approach that can be adapted to frame authentic problem-based learning.

Key to this approach is students are given the opportunity to; 1) identify the problems, 2) analyse the problems, 3) make connections between observations, concepts and theories, and 4) effectively communicate the findings and solutions, i.e. in a way that is relevant to the 'client' of the problem. This is a student managed approach with guidance provided by the mentor, and if possible the opportunity to negotiate with problems client. This is consistent with adult learning theories where personal relevance, experience and context result in deep engagement by the learner (Knowles, 1980; Lave and Wenger, 1991; Field et al., 2017). This approach has appropriated problems engaging external stakeholders including; environmental consulting companies, local councils, national and state government authorities, local landholders, and the 19th World Congress of Soil Science (Hartemink et al., 2014). This engagement illustrates the multi-disciplinary nature of the problems and is effectively guided by the framework illustrated in Table 3.

This engagement of campus-based university students with external stakeholders brings their own experiences and contexts that drive their personal relevance. Ultimately, this will be underpinned by deep soil science knowledge bring expertise to the problem, i.e. the 'know', and the provision of relevant solutions for those who need to 'know of' soil. To illustrate this Field et al. (2017) reports the development of an eBook that incorporates industry defined soil science knowledge

Table 3. Example of using a Module Based Learning Framework to guide students in engaging with authentic problem-solving.

Module 1	*Objective*: Is to develop student's ability to make relevant observations, decide on how these can be robustly measured and use this to decide on problem(s). *Example Guiding Questions*: Compare the plant growth and how can this be measured? How many plants to sample, where are they sampled, and where on the plant to sample? Remember to consider what data do you expect and how can this be analysed and presented? *Student Managed Approach*: Decided to measure plant height, yield data, planting density. Decided to sample plants in different areas of the field. *Connections*: Planting density was uniform but noted that plants grew better on darker soil and, also questioned if variation on irrigation could be an affect
Module 2	*Objective*: Is to create connections between observations and measurements *Example Guiding Questions*: What and where to measure soil? How to measure soil texture, soil water, soil nutrients? How to analyse data, to compare plant and soil measures? *Student Managed Approach*: Sample the soil surface and measure the clay %, soil water content, and the nutrients (NPK). Students learn relevant data analysis techniques *Connections*: Higher clay content was related to large soil water capacity. This in turn resulted in 'healthier' plants. Variation in CEC, pH and nutrient status should be reflected in yield.
Module 3	*Objective*: Recognise the spatial variability of plant health and soil *Example Guiding Questions*: What to use to measures soil spatial variation? What spatial soil data is available? How to analyse spatial data? *Student Managed Approach*: Use soil maps (spatially discrete) and also proximal sensed electrical conductivity surveys (EM, spatially continuous). *Connections*: Proximal sensed data is related to soil water holding which can be mapped continuously. Students started to question the need to variable rate irrigation
Module 4	*Objective*: Effective communication of results *Example Guiding Questions*: What information data is relevant? What is the one take home message? *Student Managed Approach*: Students investigate the design and content appropriate for scientific conference acceptable posters *Connections*: Students identify the challenge of succinct data representation

with multiple entry points to suit different needs and experiences of the users. The experience of developing the Soil for Grains eBook further highlighted the need to focus on multi-disciplinary approach around authentic industry scenarios where learners can make connections and consider different perspectives (Field et al., 2011; Bouma & McBratney, 2013).

4 KEY MESSAGES

The need to those with soil expertise but who also can work with other disciplines and make the necessary connections to provide contextual solutions to soil related problems is more relevant now than ever. A modern curriculum that incorporates this is timely and is framed by recognising the similarities, differences, and contributions by those who are described as knowing, know of, or are aware of soil.

There will always be the need to have a curriculum where learners can develop a deep knowledge and be recognised as those who 'know' soil. Characterised as those who have a fundamental knowledge that is the discipline of Soil Science—soil as an object of study along with knowing its attributes and processes. This group will lead the future development of the discipline and as custodians of knowledge be responsible for brokering contextually relevant solutions. Tapping into the competitive nature and giving responsibility to the aspiration for the learner to negotiate the learning opportunities is desired by this group of learners.

Those who 'know of' soil, such as agronomists, environmental consultants and policy advisors will be engaged in understanding soil's utility and its role in providing ecosystem services and valued through its natural capital. A curriculum with multiple entry points that considers the personal relevance and experiences will serve this group of learners well, and the consensus supports opportunities for authentic problem-based learning.

Although not fully explored here the acceptance of the amateur, i.e. the love of, needs to be considered and embraced as has been done by other formal society's (e.g. geological societies). Care is their connection to soil and this may be experienced through community gardens, making and admiring of soil art forms, keepers of the history and philosophy of soil, or contributing to crowd sourced data.

Of course, in the delivered curriculum the distinction between each of these learning groups is not as distinct or abrupt, but this framework will challenge the custodian of the learning environments to take time to reflect on why such learning opportunities are being created. Collectively framing the know, know of, and aware of soil will help to develop the dimension of connectivity and ensure these pursuits are relevant.

REFERENCES

Aide, M.T. 1989. Teaching critical thinking in soil classification and a soil genesis course. *Journal of Agronomy Education*. 18: 37–39.

Ala Samarapungavan, E.L.W., Bodner, G.M. 2006. Contextual epistemic development in science: A comparison of chemistry students and research chemists. *Science Education*, 90: 468–495.

Bouma, J., McBratney, A.B. 2013. Framing soils as an actor when framing wicked environmental problems. *Geoderma*, 200–201: 130–139.

Bouma, J., Van Altvorst, A.C., Eweg, R., Smeets, P.J.A.M., Van Latesteijn, H.C. 2011. The role of knowledge when wtudying innovation and the associated wicked sustainability problems in Agriculture. *Advances in Agronomy*, Academic Press, USA, 113: 285–314.

Brevik, E.C., Hartemink, A.E. 2010. Early soil knowledge and the birth and development of soil science. *Catena* 83: 23–33.

Cattle, S. 2014. The soil judging juggernaut gathers momentum. *Soil Connects*, (1): 7.

Cattle, S., Morgan, C., Levin, M., Kim, K. 2014. Soil Judging as an Instrument for Community-Building in the Discipline of Soil Science. 20th World Congress of Soil Sciences, Jeju, Korea.

Cooper, T.H. 1991. T.E.A.M. Soil judging – An experiment. *Journal of Agronomy Education*. 20: 123–125.

CPSS 2017. Certified Practicing Soil Scientist fundamentals exam. https://www.cpssaccreditation.com.au/about-us

Debenham, N., 2015. Natured Kids. http://www.naturedkids.com/

Dominati, E., Mackay, A., Rendel, J. 2017. Understanding Soils' Contribution to Ecosystem Services Provision to Inform Farm System Analysis. In. Field D.J., Morgan C.L., McBratney A.B. (Eds.) *Global Soil Security*, Springer: 207–217.

Dominati, E., Patterson, M., MacKay, A. 2010: A framework for classifying and quantifying the natural capital and ecosystem services of soils. *Ecological Economics*, 69: 1858–1868.

Feller, C., Landa, E.R., Toland, A., Wessolek, G. 2015. Case studies of soil in art. *SOIL* 1: 543–559.

Field, D.J. & Graffe, M. 2007. From experimentation to publication. Challenging student's perceptions of scientific research. Proceedings of the International Society for the Scholarship of Teaching and Learning, Sydney, NSW.

Field, D.J., Sanderson, T. 2017. Distinguishing between capability and condition. In. Field D.J., Morgan C.L., McBratney A.B. (Eds.) *Global Soil Security*. Springer: 45–52.

Field, D.J., Yates, D., Koppi, A.J., McBratney, A.B., Jarrett, L. 2017. Framing a modern context of soil science learning and teaching. *Geoderma*. 289: 117–123.

Field, D.J., Koppi, A.J., Jarrett, L., McBratney, A. 2013. Engaging employers, graduates and students to inform the future curriculum needs of soil science. Proceedings of the Australian *Conference on Science and Mathematics Education*, Australian National University, Sept 19–21: 130–135.

Field, D.J., Koppi, A.J., Jarrett, L.E., Abbott, L.K., Cattle, S.R., Grant, C.D., McBratney, A.B., Menzies, N.W., Weatherley, A.J. 2011. Soil science teaching principles. *Geoderma* 167 (68): 9–14.

Grabinger, S.R., Dunlap, J.C. 1995. Rich environments for active learning: a definition. *Association for Learning Technology Journal.* 3: 5–34.

Hartemink, A.E., Balks, M.R., Chen, Z-S., Drohan, P., Field, D.J., Krasilnikov, P., Lowe, D.J., Rabenhorst, M., van Rees, K., Schad, P., Schipper, L.A., Sonneveld, M., Walter, C. 2013. The joy of teaching soil science. *Geoderma.* 2017–2018: 1–9.

Havlin, J., Balster, N., Chapman, S., Ferris, D., Thompson, T., Smith, T. 2010. Trends in soil science education and employment. *Soil Science Society of America Journal.* 74: 1429–1432.

Janzen, H.H., Fixen, P.E., Franzluebbers, A.J., Hattey, J., Izaurralde, R.C., Ketterings, Q.M., Lobb, D.A., Schlesinger, W.H. 2011. Global prospects rooted in soil science. *Soil Science Society of America Journal*, 75: 1–8.

Jarvis, H.D., Collet, R., Wingenbach, G., Heilman, J.L., Fowler, D., 2012. Developing a foundation for constructing new curricula in soil, crop and turfgrass sciences. *Journal of Natural Resource & Life Science*, 41: 7–14.

Keesstra, S.D., Bouma, J., Wallinga, J., Tittonell, P., Smith, P., Cerdà, A., Montanarella, L., Quinton, J.N., Pachepsky, Y., van der Putten, W.H., Bardgett, R.D., Moolenaar, S., Mol, G., Jansen, B., Fresco, L. 2016. The significance of soils and soil science towards realization of the United Nations Sustainable Development Goals. *Soil.* 2: 111–118.

Kelly, T., Reid, J., Valentine, I. 2006. Enhancing the utility of science; exploring the linkages between a science provider and their end-users in New Zealand. *Australian Journal of Experimental Agriculture.* 46: 1425–1432.

Kidd, D., Webd, M., Malone, B., Minasny, B., McBratney, A. 2015. Digital soil assessment of agricultural suitability, versatility and capital in Tasmania, Australia. *Geoderma Regional* 6: 7–21.

Kitchen Garden Foundation, 2015. Stephanie Alexander Kitchen Garden Foundation. https://www.kitchengardenfoundation.org.au/.

Knowles, M., 1980. The Modern Practice of Adult Education: Andragogy Versus Pedagogy. *Cambridge Adult Education*, Englewood Cliffs, NJ.

Koch, A., McBratney, A.B., Adams, M., Field, D.J., Hill, R., Lal, R., Abbott, L., Angers, D., Baldock, J., Barbier, E., Bird, M., Bouma, J., Chenu, C., Crawford, J., Flora, C.B., Goulding, K., Grunwald, S., Jastrow, J., Lehmann, J., Lorenz, K., Minansy, B., Morgan, C., O'Donnell, A., Parton, W., Rice, C.W., Wall, D.H., Whitehead, D., Young, I., Zimmermann, M. 2013. Soil security: solving the global soil crisis. *Global Policy Journal.* 4: 434–441.

Koppi, T., Field, D.J., McBratney, A.B., Hartemink, A. 2010. The need for soil science amateurs. 19th World Congress of Soil Science, Brisbane, Australia: 30–33.

Lave, J., Wenger, E. 1991. Situated Learning. Legitimate Peripheral Participation University of Cambridge Press, Cambridge.

MacEwan, R.J., MacEwan, A.S.A., Toland, A.R. 2017. Engendering Connectivity to Soil through Aesthetics. In. Field, D.J., Morgan, C.L., McBratney, A.B. (Eds.) *Global Soil Security.* Springer. 351–363.

Malone, B.P., McBratney, A.B., Minasny, B. 2013. Spatial scaling for digital soil mapping. *Soil Science Society of America Journal.* 77: 890–902.

McBratney, A.B., Field D.J., Koch, A. 2014. The dimensions of Soil Security. *Geoderma*, 213: 203–213.

McBratney, A.B., Field, A.J., Jarrett, L. 2017. General concepts of valuing and caring for soil. In. Field, D.J., Morgan, C.L., McBratney, A.B. (Eds.) *Global Soil Security.* Springer. 101–108.

McBratney, A.B., Field, D.J. 2015. Securing our soil. *Soil Science and Plant Nutrition.* 61: 587–591.

McBratney, A.B., Morgan C.L.S., Jarret L. 2017. The Value of Soil's Contributions to Ecosystem Services. Field, D.J., Morgan, C.L., McBratney, A.B. (Eds.) *Global Soil Security.* Springer. 227–235.

Mims, F.M.III 1999. Amateur Science–Strong Tradition, Bright Future. Essays on Science and Society, *Science*, 248: 55–56.

Morgan, C.L.S., Morgan, G.D., Bagnall, D. 2017. Social license to secure soil. In. Field, D.J., Morgan, C.L., McBratney, A.B. (Eds.) *Global Soil Security.* Springer. 247–251.

Napier, T. 1998. Soil and water conservation policy approaches in North America, Europe, and Australia. Water Policy, 1: 551–565.

NERC, 2012. Most Wanted Postgraduate and Professional Skill Needed in the Environment Sector. National Environment Research Council, Swindon, UK.

Nowotny, H., Scott, P., Gibbons, M. 2002. Re-thinking Science. Knowledge and the Public in an Age of Uncertainty Polity Press, Cambridge, UK, 1–21.

Robinson, D.A. 2012. Natural capital, ecosystem services and soil change: why soil science must embrace ecosystem service approach. *Vadose Zone Journal.* 11.

Robinson, D.A., Fraser, I., Dominanti, J., Davidsdóttir, B., Jónsson, J.O.G., Jones, L., Jones, S.B., Tuller, M., Lebron, I., Bristow, K.L., Souzai, D.M., Banwart, S., Clothier, B.E. 2014. On the Value of Soil Resources in the Context of Natural Capital and Ecosystem Service Delivery. *Soil Science Society of America.* 78: 685–700.

Rolfe, G., Freshwater, D. and Jasper, M. (2001). Critical reflection in nursing and the helping professions: a user's guide. Basingstoke: Palgrave Macmillan

Shelley, W., Lawlay, R., Robinson, D.A. 2013. Technology: crowd-sourced soil data for Europe. *Nature* 496–300.

Sojka, R.E., Upchurch, D.R., Borlaug, N.E. 2003. Quality soil management or soil quality management: performance versus semantics. *Advances in Agronomy.* 79: 1–68.

SSSA, 2007. Soil science professional practical exam Performance Objectives. *Soil Science Society of America's Council of Soil Science Examiners.* https://www.soils.org/files/certifications/practice-exam-objectives.pdf

SSSA, 2017. Soil Judging, *Soil Science Society of America.* https://www.soils.org/undergrads/soils-contest

Stockmann, U., Adams, M., Crawford, J.W., Field, D.J., Henakaarchchi, N., Jenkins, M., Minasny, B., de Courcelles, V.R., Singh, K., Wheeler, I., Abbott, L., Angers, D., Baldock, J., Bird, M., Brookes, P.C., Chenu, C., Jastrow, J., Lal, R., Lehmann, C.J., O'Donnell, A.G., Parton, W.J., Whitehead, D. Zimmermann, M. 2013. The knowns, known unknowns and unknowns of sequestration of soil organic carbon. *Agriculture, Ecosystems & Environment* 164: 80–99.

Toland, A.R., Wessolek, G. 2010. Core Samples of the Sublime On the Aesthetics of Dirt. In: Landa, E.R., Feller, C. (Eds.), *Soil and Culture*. Springer, London, New York, 239–257.

Tugel, A.J. 2005. Soil change, soil survey, and natural resource decision making. *Soil Science Society of America Journal*. 69: 783–747.

Viscarra-Rossel, R.A, Walvoort, D.J.J., McBratney, A.B., Janik, L.J., Skjemstad, J.O. 2006. Visable, near infrared, mid infrared, or combined diffuse reflectance spectroscopy for simultaneous assessment of various soil properties. *Geoderma*. 131: 59–75.

White, R., Stoecklin, V.L. 2008. Nurturing Children's Biophilia: Developmentally Appropriate Environmental Education for Young Children. White Hutchinson Leisure & Learning Group, https://www.whitehutchinson.com/children/articles/nurturing.shtml

Yaalon, D. 1996. Soil science in transition: soil awareness and soil care research strategies. *Soil Science*. 161: 3–8.

The non-anthropocentric value of soil and its role in soil security and Agenda 2030

P.E. Back, A. Enell & Y. Ohlsson
Swedish Geotechnical Institute, Linköping, Sweden

ABSTRACT: Soil Security is basically an anthropocentric concept, i.e. it is mainly based on sustainability from a human perspective. Human needs are also the starting point for the 2030 Agenda for Sustainable Development, but several of these goals also address non-anthropocentric aspects. This paper discusses the role of non-anthropocentric value in the Soil Security framework for fulfilment of Sustainable Development Goals (SDGs) in Agenda 2030, especially the intrinsic value of soil. This ethical value is independent on whether or not soil is of benefit to man. We explore the extent to which non-anthropocentric aspects are reflected in the SDGs and where in the Soil Security framework these aspects could be addressed. The conclusion is that the framework does have the potential to meet the Agenda 2030 SDG requirements with respect to the non-anthropocentric value of soil, with some modifications and clarifications. This will further enhance the framework's applicability.

1 INTRODUCTION

Soil has always been of fundamental importance to mankind. We benefit from soil in many different ways, both directly and indirectly. Soil delivers a multitude of goods and services to mankind and many of these are included in the concept *ecosystem services* (MEA, 2005). Soil is considered a *natural capital*, generating a flow of ecosystem services (Dominati et al., 2010). The aggregated value of these ecosystem services is often considered to be the value of the soil ecosystem. But does it comprise the complete set of values of the soil ecosystem that needs to be considered? There has been concern that something is missing when nature's value is appreciated only by what it can deliver to humans (Batavia & Nelson, 2017). A common argument is that nature also has a value on its own, regardless of actual or potential goods and services it can deliver to humans. This way of reasoning is founded on an ethical view that is non-anthropocentric, i.e. an ecosystem has its own right to exist regardless of any human interest (Hågvar, 1998).

Soil Security, as defined by McBratney et al. (2014), is basically an anthropocentric concept, i.e. the focus is on soil from a human perspective. They state that "*...soil security is a concept of securing soil for the sustainable development of humanity...*". It is a complement to the six global challenges Food Security, Water Security, Energy Security, Climate Change Abatement, Biodiversity Protection and Ecosystem Service Delivery, which all focus on servicing mankind. They all have biophysical attributes but also include economic, social and policy aspects.

In 2015, the 17 Sustainable Development Goals (SDGs) of the 2030 Agenda for Sustainable Development were adopted by world leaders (United Nations, 2015). These sustainability goals are also focusing on human interests but some non-anthropocentric aspects are also addressed.

A legitimate question is whether it matters if the value of soil is considered from an anthropocentric or non-anthropocentric point of view, in order to reach Soil Security and meet the Agenda 2030 SDGs. If the soil has a high value from an anthropocentric view, and for this reason should be protected, does this imply that the non-anthropocentric values are protected implicitly? We believe that this will be the situation in many cases, but not always. There may be at least three situations when it is not sufficient to only consider the soils anthropocentric value: (1) when people benefit very little from the soil, (2) when the soil's non-anthropocentric value is very important, and (3) when a complete description of the soil's different values is warranted for other reasons. In all these cases, it will not be sufficient to only consider the anthropocentric value of soil. In addition, the complementary issue of soil's intrinsic value must be addressed, i.e. the value of the soil ecosystem in its own, independent of the human benefit. Examples of the first case above are soil at many contaminated sites or at remote locations with hardly any people. The second case encompasses soil in sensitive ecosystems, nature preserves etc.

The third situation corresponds to educational purposes or to illustrate constraints in anthropocentric approaches based on ecosystem services.

In this paper we want to illuminate the ethical aspect regarding soil, specifically the non-anthropocentric value with its base in environmental ethics, here applied to the soil ecosystem as part of nature. The objective of the paper is to illustrate how the non-anthropocentric value of soil can be considered within the Soil Security framework, coupled to the Agenda 2030 for sustainable development. The paper is written from a contaminated land perspective but the content is general.

2 ENVIRONMENTAL ETHICS

An important ethical question is whether the soil ecosystem has a value beyond what it provides humans. In order to answer this question, a few basic terms and concepts must be briefly explained.

Environmental ethics developed relatively late, as a result of the environmental problems that became apparent during the 1960s and 1970s, although attempts were made even earlier by e.g. Leopold (1949). A central issue in environmental ethics is to explain which objects or entities in nature having moral status, i.e. if and to what extent the objects must be considered for their own sake. Objects having moral status are considered to have *intrinsic value*, i.e. a value of their own (sometimes denoted inherent value or final value). This value is fundamentally different from *instrumental value*, which is the value an entity has for something else (Brennan & Lo, 2016). Nature's instrumental value for humans is often expressed as ecosystem services (Batavia & Nelson, 2017). The instrumental value of soil is the central issue in the Capital domain in the Soil Security framework. The intrinsic value of soil is more abstract and has its base in ethics.

A number of environmental ethical views have been presented, and the most important ones with respect to soil will be briefly described. *Anthropocentrism* is a human-centered view where human needs are at its core and nature only has a value as a resource to fill human needs (Batavia & Nelson, 2017). According to this view, man is the only creature with an intrinsic value. Historically, the view of nature in western civilization has almost entirely been anthropocentric. The Brundtland Commission (WCED, 1987) propagated for a modern variant of anthropocentrism, the so-called *intergenerational anthropocentrism*, or the ethics of sustainable development, in which not only people in the present generation are considered as moral objects but also people in future generations. With this view, soil has a value based of its benefit to people in both the present and future generations. However, non-humans' benefit from soil is still not considered.

In *biocentrism*, the group of objects with intrinsic value is extensively increased. All living creatures, and only those, have moral status and consequently intrinsic value (Brennan & Lo, 2016). However, only individual organisms are considered to have intrinsic value. Larger entities, such as habitats, ecosystems, rivers, or non-living parts of ecosystems are not valued intrinsically by biocentrists. Different variants of biocentrism have developed, for example sentientism where the focus is on creatures with the capacity to experience pain. A biocentric view of soil would imply that only the organisms in the soil have value, not the abiotic soil material or the soil as a system. In the most extreme biocentric view, all organisms have the same intrinsic value, which of course is problematic in reality.

In the environmental debate it soon became apparent that a biocentric approach was not sufficient to handle complex systems in nature, like ecosystems. Consequently, ecocentrism developed. Note that the term biocentrism is sometimes used in a broad sense that includes ecocentrism. It is important to distinguish between strong ecocentrism and weak ecocentrism, as they are fundamentally different from a human perspective. In strong ecocentrism, ecosystems have stronger intrinsic value than humans. This resulted in a debate about 'eco fascism', because humans were considered subordinate to ecosystems (Callicott, 1999). As a result, weak ecocentrism was developed, where human interests are considered stronger. Both ecological entities (populations, species, habitats and ecosystems) and the individual members involved in these entities (humans, animals and plants) have a value, but the intrinsic value of humans is the strongest. Even abiotic components like minerals and water have some value according to this view, because they are part of the ecosystem (Rolston, 1988). According to weak ecocentrism, soil has intrinsic value as a system. Even the individual components of a soil ecosystem have value, although weaker than the value of the system as a whole (Rolston, 1988).

Although most environmental philosophers (but not all; see e.g. Norton, 1995) conclude that nature has intrinsic value, there is no consensus on the very essence of this type of value. Rolston (1988) and Callicott (1999) represent two important, but different views on this issue. Rolston argues that nature's intrinsic value is an inherent property that is objective and independent of human beliefs. This view is also supported by the philosopher Naess (1993). Callicott on the other hand, claims that humans assign intrinsic value to a nonhuman entity for its own sake, i.e. the value does not necessarily exist in the absence of human valuers (Batavia & Nelson, 2017). In this paper we will not propagate for one view over the other, but we note that Callicott's view is more practical for real-world problems because it allows us to assign intrinsic value subjectively, rather than trying to estimate an objective but often uncertain value.

There are numerous different definitions of intrinsic value based on a variety of philosophical views; see reviews in e.g. Vilkka (1997) and Batavia & Nelson (2017). One possibility is to consider intrinsic value in terms of strength rather than magnitude. A strong intrinsic value indicates that the soil ecosystem has a strong 'right' to be spared from unnecessary harm. This should not be confused with an absolute right to be protected or that intrinsic value is infinite (MEA, 2005; Vucetich et al., 2015). Strong human interest may outweigh even a strong intrinsic value, at least according to weak ecocentrism.

3 THE NON-ANTHROPOCENTRIC VALUE OF SOIL

The value of soil can be formulated based on the definitions above, as illustrated in Figure 1. For a more detailed description we refer to Back et al. (2016). The illustration is based on a combination of anthropocentric and ecocentric ethics, which form the basis for the Swedish Environmental Code (Hansson, 2014). The presentation below is based on this Swedish perspective but its application is general.

As stated earlier, the anthropocentric value of soil corresponds to the soil's instrumental value to humans, commonly referred to as ecosystem services. The instrumental value of soil is the central issue in the Capital dimension in the Soil Security framework (McBratney et al., 2014).

The non-anthropocentric value consists of two parts; instrumental value and intrinsic value. The instrumental value is the benefit the soil delivers to other parts of the ecosystem, such as food for birds etc. (in this case excluding human benefit). This type of value is discussed by e.g. Rolston (1988) and Fosci & West (2016). It is sometimes referred to as an 'ecological service' (Hågvar, 1998). The focus of non-anthropocentric value of soil is generally on the intrinsic value component.

Intrinsic value of soil is not a new concept in soil science; see e.g. Warkentin (1995) and Hågvar (1998). However, detailed explanations of the very nature of this type of value should be derived from the ethics literature. Subjective intrinsic value of soil, according to Callicott's view, is anthropogenic (attributed *by* humans) but non-anthropocentric (i.e. not attributed *to* humans). Objective intrinsic value, according to Rolston's view, is neither anthropogenic nor anthropocentric.

Note that Figure 1 does not necessarily cover all possible aspects of the value of soil. For example, Pascual et al. (2017) consider relational values and Rolston (1988) advocates systemic values. A distinction can also be made between intrinsic and extrinsic values, where extrinsic value arises out of the relation with some external factor. The subjective intrinsic value in Figure 1 can actually be regarded as a type of extrinsic value (Batavia & Nelson, 2017) because a human is involved in its assignment, according to Callicott (1999). We use the term intrinsic value in a manner that is common in the literature, although not always philosophically precise.

Despite its simplicity, Figure 1 is probably sufficient for most situations where the value of soil should be considered in order to achieve Soil Security and the Agenda 2030 SDGs. How the non-anthropocentric values in Figure 1 can be addressed in the Soil Security framework is discussed below.

4 APPLICATION TO SOIL SECURITY

4.1 *Sustainable development goals*

The 17 SDGs of the 2030 Agenda for Sustainable Development (United Nations, 2015) were adopted by world leaders in 2015 and came into force on 1 January 2016. As noted earlier, the concept of sustainable development has its base in intergenerational anthropocentrism, where focus is on the human needs, both today and in the future. However, the SDGs also cover aspects that to some extent are non-anthropocentric. In preparing this paper, a systematic review of all SDGs was carried out with the purpose of identifying SDG subgoals that in one way or the other refer to soil's non-anthropocentric value. The conclusion is that primarily SDG No 15 (Sustainably manage forests, combat desertification, halt and re *verse land degradation, halt biodiversity loss*) contains such subgoals. The six most relevant subgoals of SDG No 15 are listed in short in Table 1, together with a notion of in which Soil Security dimension work is required in order to secure soil and meet the goals.

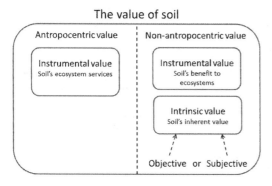

Figure 1. Schematic illustration of the value of soil.

Table 1. Identified subgoals in Agenda 2030 with relevance for non-anthropocentric aspects (N), and a notation of where in the Soil Security framework they need to be addressed. The need for addressing anthropocentric aspects (A) to secure soil and meet the subgoals is indicated for comparative purpose. Letters within parentheses indicate aspects that *probably* need to be addressed.

No	Goal 15 targets	Capability	Condition	Capital	Connectivity	Codification
15.1	... ensure the conservation, restoration and sustainable use of terrestrial and inland freshwater ecosystems and their services...	A, N	A, N	A	(A, N)	(A, N)
15.3	... combat desertification, restore degraded land and soil...	A, N	A, N	(A)	(A, N)	(A, N)
15.4	... ensure the conservation of mountain ecosystems, including their biodiversity, in order to enhance their capacity to provide benefits that are essential for sustainable development.	A, N	A, N	A	(A, N)	(A, N)
15.5	... reduce the degradation of natural habitats, halt the loss of biodiversity and, by 2020, protect and prevent the extinction of threatened species.	A, N	A, N		(A, N)	(A, N)
15.9	...integrate ecosystem and biodiversity values into national and local planning, development processes, poverty reduction strategies and accounts.			A	A, N	A, N
15.10	Mobilize and significantly increase financial resources from all sources to conserve and sustainably use biodiversity and ecosystems.				A, N	A, N

Table 1 clearly shows that if the Soil Security concept is to be used for fulfilment of the SDGs, non-anthropocentric aspects need to be integrated in the framework. Non-anthropocentric aspects relate to four of the five dimensions of the present Soil Security framework. The only exception is Capital. Although this dimension concerns values, the anthropocentric focus is inherently manifested by the common definition of the term capital. Therefore, at present, there is no explicit dimension where the non-anthropocentric value of soil clearly fits in the framework. There are, however, several options to further develop the Soil Security concept in order to incorporate non-anthropocentric aspects. Hence, the five Soil Security dimensions are further discussed below regarding their possibility to consider non-anthropocentric aspects.

4.2 *Dimension 1 & 2: Capability and condition*

Capability can be defined as the optimal state of the soil, i.e. the reference state of the soil (McBratney et al., 2014). It reflects the soil's inherent properties. Warkentin (1995) related intrinsic value of soil to soil quality. This can be compared to Rolston's view of objective intrinsic value, which he means is a consequence of complex organisms and systems in nature. Thus, such entities have intrinsic value (or systemic value) *per se*, according to Rolston (1988). Although intrinsic value is not quantifiable like physical soil parameters it is related to the soil's Capability. However, the relation is not simple. A high-yielding soil does not necessarily have a higher intrinsic value than a low-yielding. The intrinsic value is more related to biodiversity, uniqueness, irreplaceability, extremeness, and fragility of the soil ecosystem; see e.g. Warkentin (1995) and Hopkins & Gregorich (2013). Other criteria for the strength of soil's intrinsic value can probably be derived from ecocentrism. One such additional criterion could be complexity of the soil ecosystem (Rolston, 1988).

The soil's Condition can be defined as the current state of the soil, which can deviate from its capability. An important question is: Has the soil lost some of its intrinsic value when human activity has negatively affected the condition of the soil compared to its reference state (Capability)? Taking contaminated soil as an example, the question can be reformulated: Does contaminated soil have a weaker intrinsic value than a similar but uncontaminated soil? The answer is neither self-evident nor simple. According to an ecocentric ethics the answer is 'no'. The intrinsic value is coupled to the soil ecosystem as a moral object. A moral object does not lose its value if the object is harmed in some way. For example, a human's intrinsic value is not reduced if a leg is broken. On the other hand, the intrinsic value of humans is generally considered to be much stronger than for soil, so the comparison is, of course, somewhat halting. In addition, some human actions result in irreversible alteration of the soil, reducing or changing its condition indefinitely. Should this change the reference state to the phenoform in the Capability dimension? Such a change can be problematic since it opens up for circular reasoning: Assume that the soil is contaminated, which changes the reference state from the genoform to the phenoform. When only the phenoform is considered there is no need for remediation (if this action only is triggered by preserving the soils intrinsic value, which may be the case). The problem is accentuated when the reference state is changed to the phenoform at an early stage of a risk assessment process. At this stage, the risks are not fully known and the phenoform could constitute a risk. Changing the reference state in this manner could counteract sustainable solutions. To avoid this fallacy, great care should be taken before the reference state is changed for contaminated soil.

It is important to distinguish between the contaminated soil and the soil as a system. Removal of contaminated soil does not necessarily mean that the intrinsic value is lost. The intrinsic value is strongest for the soil as a system, which implies that replacement of contaminated soil with similar uncontaminated soil is ethically justifiable.

Another problem in this context is how foreign material in the ground should be regarded. It is not uncommon that filling, construction waste, ash and similar materials that may contain contaminants are present in the ground, separately or in a mix with natural soil material. In such cases, arguments are sometimes made that these materials lack intrinsic value and capability to support organisms, and hence, there is no need to remediate the soil. This relatively common misconception does not take into account that it is the soil as an ecosystem that has intrinsic value. The fact that individual volumes of material lack intrinsic value (and capability) cannot be used as an argument to avoid dealing with the problem.

All this illustrates that capability (including its reference states), condition, and intrinsic value are intimately associated with each other.

4.3 Dimension 3: Capital

Capital is almost always considered an anthropocentric concept. This implies that the Capital dimension generally includes anthropocentric instrumental values (Figure 1), often expressed by *ecosystem services* or as *natural capital*. However, attempts have been made to couple the intrinsic value of nature to the Capital dimension. For example, *existence value* has been used as a monetary measure of intrinsic value; see Batavia & Nelson (2017) for a discussion. The existence value of soil captures the value that people derive from simply knowing that the soil exists, i.e. a form of ecosystem service that partially overlaps intrinsic value (MEA, 2005). Although existence value is closely related to intrinsic value, it is an instrumental value (MEA, 2005; Batavia & Nelson, 2017) with an anthropocentric focus. It does not capture soil's value for its own sake, as a soil ecosystem. Vucetich et al. (2015) do not categorically reject that intrinsic value can be subject to economic valuation, but they mean that the question is wrong. A more appropriate question would be: What is the best way to handle competing values that involve intrinsic values? They conclude that few people would agree that that economic valuation is the best tool for this. Batavia & Nelson (2017) argue that it is perfectly consistent to recognize both instrumental and intrinsic value in an entity. They ask that perhaps the ecosystem services framework is meant to complement, rather than replace, the idea of nature's intrinsic value.

Another approach is to try to include the idea of intrinsic value in the concept natural capital. For example, Lawn (2000) has built an economic framework for sustainable development to include a biocentric view, stating that natural capital includes both instrumental and intrinsic value. He derives the intrinsic value from the uniqueness and sentience of natural capital. His conclusion is that the optimal macroeconomy will usually be smaller when instrumental and intrinsic value of natural capital is combined, compared to if a purely anthropocentric approach is applied, where only instrumental value is considered. Thus, intrinsic value of natural capital can act as a restriction which counteracts excessive and unsustainable use of natural capital.

Baveye et al. (2016) observe that viewing soils as natural capital may be helpful, because this capital is unfathomably large in the case of soils. This may encourage soils to be managed in the same way as priceless treasures like historical and cultural sites. Thus, natural capital comprises more than monetary value. However, the 'cultural capital' discussed by Baveye et al. (2016) is not identical to intrinsic value, rather existence value.

Soil biodiversity can also be considered part of the natural capital of soil (McBratney et al. (2017). It is closely related to intrinsic value (United Nations, 1992; Haines-Young & Potschin, 2010). However, the value of biodiversity is commonly expressed by ecosystem services in the Capital dimension, not as an intrinsic value.

As indicated, including the non-anthropocentric value of soil in the Capital dimension of Soil Security is not trivial. Including intrinsic value in ecosystem services and natural capital can be questioned. It assumes that a non-anthropocentric concept can be incorporated in anthropocentric concepts, which of course is problematic and somewhat confusing. Some modification in the definition of the Capital dimension is probably required in order to fully integrate the nonhuman value of soil.

4.4 *Dimension 4: Connectivity*

The Connectivity dimension brings in the social aspects of soil. It relates to the knowledge and resources we have for soil management, decisions, and how and for what we value the soil. McBratney et al. (2014) point out that *"if there is no connectivity to the soil then the soil itself may not be valued and is prone to not being managed to its best condition"*. Furthermore, soil is a nonrenewable resource, and hence, there is need for intergenerational stewardship with knowledge transfer between generations. The authors also argue that adoption of the precautionary principle may be necessary in order to secure soil resources that we still not have knowledge about. These social aspects are all anthropocentric, but McBratney et al. (2014) do point out that *"it* [connectivity] *raises the question regarding the need for a soil ethic and in doing so whether soil should only be valued for the well-being of humans"*.

So, is it possible to expand the Connectivity dimension to also include non-anthropocentric values and aspects? Knowledge and education strategies, as well as lobbyism, would certainly be important for this matter, but the essential issue is if we are willing to acknowledge the non-anthropocentric value of soil. A central aspect to elucidate is how people think about soil that they do not benefit from themselves. A number of social studies have been made about how people in some western countries regard nature; see references and summaries in Naess (1993), Lundmark (2000) and Vucetich et al. (2015). The results were relatively consistent: A great majority of the respondents considered that nature has intrinsic value. In addition, MEA (2005) discuss several bases for intrinsic value in non-western religious and cultural worldviews, where concepts with similarities to intrinsic value are common. It thus seems to be common belief that nature has intrinsic value. Consequently, acknowledging the intrinsic value of soil does not seem farfetched.

4.5 *Dimension 5: Codification*

Codification is the dimension of policy and legislation. This is a very important dimension for the non-anthropocentric value of soil. Policy and legislation will very much control if the intrinsic value of soil is considered or not. One of the most important international legal instruments regarding intrinsic value is probably the Convention of Biological Diversity (United Nations, 1992), which already in its first sentence acknowledges the intrinsic value of biodiversity. Although international environmental law has encompassed intrinsic value to some extent (Fosci & West, 2016), the intrinsic value of soil is generally not acknowledged in national legislation. An example of a national legislation, that to some extent acknowledges nature's intrinsic value, is the Swedish Environmental Code. It states that nature has a value on its own. This implies that there is a rather strong foundation for the intrinsic value of soil ecosystems in Sweden, although soil is not implicitly mentioned in the code. However, in the Swedish national guidelines for contaminated land, the soil ecosystem is considered a protection target, regardless of its benefit to humans (Back et al., 2016). Thus, the soil's non-anthropocentric value is implicitly considered. Examples of other legislation that acknowledges the intrinsic value of nature are discussed by e.g. Filgueira & Mason (2009) and Fosci & West (2016).

5 DISCUSSION

Although the main focus of the Soil Security framework is on human needs, this paper has shown that it is important to also include non-anthropocentric aspects in the framework. The concept of intrinsic value is not merely philosophical; it can have major practical consequences, as indicated in this paper. Intrinsic value is a concept of importance in environmental ethics, widely accepted among e.g. conservation biologists (Piccolo, 2017), and also recognized by Millennium Ecosystem Assessment (MEA, 2005). In addition, some support for the concept can be found in both legislation and international conventions (Fosci and West, 2016). By taking both instrumental and intrinsic values of soil into account, the possibilities of achieving Soil Security and meeting Agenda 2030 SDGs will most likely increase.

Consideration of the non-anthropocentric value of soil can be argued for in several ways. First, international environmental law and policy frameworks (codification) that acknowledge nature's intrinsic value do already exist. However, the codification regarding soil's non-anthropocentric value is generally rather weak. A second argument is that ecosystem services may not always deliver a sufficient motive for protection of soil. When the ecosystem services approach is used, it is generally assumed that the value of these goods and services is sufficiently high to warrant protection. This is not always the case, especially not for small parcels of land, or for soil in sparsely populated areas. A third argument is ethical: Soil ecosystems have a right to be protected from unnecessary and harmful actions, at least from an ecocentric perspective. This ethic emphasizes the soil ecosystem as a moral object.

In the following three cases, ignoring non-anthropocentric values is especially problematic:

(1) when ecosystem services are difficult to assess or are of low value, (2) when the intrinsic value of soil is especially strong, and (3) when human activity has caused negative effects on soil organisms or on other parts of the ecosystem, e.g. secondary poisoning in the food web. In these situations, ignoring the non-anthropocentric value of soil could counteract sustainability and cause threats to persist.

Although it is far from simple to address the nonhuman value of soil in practice, we still believe that the Soil Security framework has the potential to meet the Agenda 2030 SDG requirements with respect to this type of value. However, this will require some modifications and clarifications to further enhance the framework's applicability, as described in the previous sections. There is currently no single scientifically correct or generally accepted way of expressing the value of soil, and sometimes terms are used that deviate from the terminology in this paper. Relational values, ecological services, biodiversity values etc. are frequently encountered in the literature. Therefore, a pluralistic approach to value of soil would be advisable, keeping the framework open to different perspectives.

REFERENCES

Back, P., Hermansson, S., Rosén, L., Volchko, Y., Wiberg, K., Fransson, M. & Enell, A. 2016. Does soil have value beyond what it provides humans? In Proceedings of the 2nd International Conference of *Global Soil Security*, Paris, December 5–6, 2016.

Batavia, C. & Nelson, M.P. 2017. For goodness sake! What is intrinsic value and why should we care? *Biological Conservation* 209(2017): 366–376.

Baveye, P., Baveye, J. & Gowdy, J. 2016. Soils "ecosystem" services and natural capital: Critical appraisal of research on uncertain ground. Frontiers in *Environmental Sciences* June 2016(4); 41: 1–49.

Brennan, A. & Lo, Y. 2016. Environmental Ethics. *The Stanford Encyclopedia of Philosophy* (Winter 2016 Edition), E.N. Zalta (ed).

Callicott, J.B. 1999. *Beyond the land ethic. More essays in environmental philosophy*. State University of New York Press.

Dominati, E., Patterson, M., & Mackay, A. 2010. A framework for classifying and quantifying the natural capital and ecosystem services of soils. *Ecological Economics* 69(2010): 1858–1868.

Filgueira, B. & Mason, I. 2009. *Wild Law: Is there any evidence of earth jurisprudence in existing law and practice?* London: UK Environmental Law Association and the Gaia Foundation.

Fosci, M. & West, T. 2016. In whose interest? Instrumental and intrinsic value in biodiversity law. In M. Bowman, P. Davies & E. Goodwin (eds), *Research handbook on biodiversity and law*: 55–77. University of Nottingham.

Hansson, F. 2014. "…to promote sustainable development…" The expression "sustainable development" in the Swedish Environmental Code as seen from the perspective of environmental ethics. Graduate Thesis. Faculty of law, Lund University.

Haines-Young, R. & Potschin, M. 2010. The link between biodiversity, ecosystem services and human well-being. In D.G. Raffaelli & C.L.J. Frid (eds), *A new synthesis*: 110–139. Cambridge University Press.

Hopkins, D.W. & Gregorich, E.G. 2013. Managing the soil-plant system for the delivery of ecosystem services. In P. Gregory & S. Nortcliff (eds), *Soil conditions and plant growth*: 390–416. Chichester: Wiley-Blackwell.

Hågvar, S. 1998. The relevance of the Rio-convention on biodiversity to conserving the biodiversity of soils. Applied Soil Ecology 9(1998): 1–7.

Lawn, P.A. 2000. *Toward sustainable development: An ecological economics approach*. Boca Raton: Lewis Publishers.

Leopold, A. 1949. *A sand county almanac and sketches here and there*. Oxford: Oxford University Press.

Lundmark, F. 2000. *Människan i centrum. En studie av antropocentrisk värdegemenskap*. Doktorsavhandling, Uppsala universitet. Förlags ab Gondolin. (PhD Thesis, In Swedish)

McBratney, A., Field, D. & Koch, A. 2014. The dimensions of soil security. *Geoderma* 213(2014): 203–213.

McBratney, A., Field, D. & Morgan, C. 2017. Soil security: A rationale. In D. Field, C. Morgan & A. McBratney (eds), *Global soil security*: 3–14. Springer.

Millennium Ecosystem Assessment (MEA) 2005. *Ecosystems and human well-being: A framework for assessment*. Chapter 6: Concepts of ecosystem value and valuation approaches: 127–147. Island Press.

Naess, A. 1993. Intrinsic value: Will the defenders of nature please rise. In P. Reed & D. Rothenberg (eds), *Wisdom in the open air*: 70–82. Minneapolis: University of Minnesota Press.

Norton, B.G. 1995. Why I am not a nonanthropocentrist: Callicott and the failure of monistic inherentism. *Environmental Ethics* 17(4): 341–358.

Pascual, U., Balvanera, P., Díaz, S., Pataki, G., Roth, E., Stenseke, M., Watson, R.T. et al. 2017. Valuing nature's contributions to people: the IPBES approach. *Current Opinion in Environmental Sustainability* 2017, 26: 7–16.

Piccolo, J.J. 2017. Intrinsic values in nature: Objective good or simply half of an unhelpful dichotomy? *Journal for Nature Conservation* 37(2017): 8–11.

Rolston, H. 1988. *Environmental Ethics: duties to and values in the natural world*. Philadelphia: Template University Press.

United Nations 1992. *Convention on Biological Diversity*, 1760 UNTS 79.

United Nations 2015. *Transforming our world: the 2030 Agenda for Sustainable Development*. A/RES/70/1.

Vilkka, L. 1997. The intrinsic value of nature. *Value Inquiry Book Series*, Volume 59. Amsterdam: Rodopi.

Vucetich, J.A., Bruskotter, J.T. & Nelson, M.P. 2015. Evaluating whether nature's intrinsic value is an axiom of or anathema to conservation. *Conservation Biology* 29(2): 321–332.

Warkentin, B.P. 1995. The changing concept of soil quality. *Journal of Soil and Water Conservation* May-June: 226–228.

World Commission on Environment and Development (WCED) 1987. *Our Common Future*. United Nations.

Soil awareness in Italian high schools: A survey to understand soil knowledge and perception among students

M.C. Moscatelli & S. Marinari
Department for Innovation in Biological, Agrofood and Forest Systems (DIBAF), University of Tuscia, Viterbo, Italy

S. Franco
Department of Economics, Engineering, Society and Business Organization (DEIM), University of Tuscia, Viterbo, Italy

ABSTRACT: Connectivity is one of the five dimensions framing the concept of soil security. It is a social dimension aimed to promote soil knowledge and awareness through communication and education. This paper reports the results of a case study based on a survey among 500 Italian high school students with the aim to assess knowledge of soil properties, perception of its functions and opinion on causes and consequences of its degradation. The level of knowledge of soil basic properties was quite low with only 39% of correct answers. The regression analysis between soil knowledge and the socio-demographic data revealed its significant dependence from the educational background pointing to instruction as the predominant factor underlying soil knowledge. However, a positive correlation was found between soil knowledge and perception of soil functions indicating that education is the essential pre-requisite for soil awareness. Soil degradation was mostly perceived when linked to other environmental issues such as urbanization, waste disposal (as causes), water and air pollution (as consequences). Students' opinion on effective tools to raise soil awareness and promote soil culture was mainly related to individual attitudes and to digital social behaviors. These last aspects should be considered for policies aimed to effectively connecting with youngsters to promote and build up a solid soil culture within modern society.

1 INTRODUCTION

The recently introduced concept of soil security is characterized by a wider and more integrative view of the soil resource, bringing new perspectives that are necessary for soil protection (McBratney and Field, 2015; McBratney et al., 2014; Bouma et al., 2012). Connectivity is one of the five dimensions that frame this new concept and that should be assessed to secure soil. In particular, this new social dimension places soil at the centre of communication and teaching, claiming for appropriate soil education approaches and communication tools that can raise soil knowledge and awareness while connecting scientists, land managers and the public opinion.

However, this dimension is the least explored among the five characterizing the soil security concept (McBratney, 2014). Therefore soil connectivity needs particular attention and future work. Among the different causes that lead to disregarding soil protection within top environmental issues, cultural and educational aspects should be considered (Moscatelli et al., 2011; Hermann, 2006).

Individual perception of the diverse environmental issues is considerable when the effects are directly experienced by man, as the case of air pollution, desertification or loss of aboveground biodiversity. Communicating the importance of the soil resource is hampered by several factors linked to its intrinsic nature: i.e. a resource hidden below the surface, a location of not easily recognizable forms of life, an idea culturally associated to dirt, death and burials (Moscatelli et al., 2011). Serious knowledge deficiencies over basic soil properties and functions may still be found within the general public, among youngsters and, alarmingly, among stakeholders and policy makers.

Enhancing awareness by communicating information on soil across the science-practice interface is also extremely important as well as sharing experiences with various citizen groups. Moreover, it may represent an effective mechanism to mobilize the policy arena (Ingram et al., 2016; Bouma et al., 2012).

In particular, exploring the level of soil awareness among youngsters is a crucial starting point for investing in the cultural asset of future generations.

The educational environment is, in fact, the milieu where attitudes and behaviours are shaped. This will help to build up positive perceptions about the concept of soil and, more in general, of natural resources (Albayrak and Hackverdi Can, 2012).

Moving from these considerations, the aim of the study is to assess the level of soil knowledge and perception among youngsters, taking as sample a group of high school students in the town of Viterbo (Central Italy).

2 SURVEY, SAMPLE CHARACTERISTICS

2.1 Survey description

The study refers to a survey performed at the University of Tuscia, within the framework of World Soil Day celebrations that take place worldwide each year in December.

The two daily events entitled "Soil: an essential resource for human and environmental health" and "To know soil, to manage soil, to create new cultural pathways to increase soil awareness" were aimed to communicate and spread a "soil culture". In particular, the target was to raise students' awareness on the factors threatening soil resource and endangering its precious functions. Each day was structured in parallel sessions allowing the active participation of a great number of students to keynote lectures, screening of short movies, practical laboratory activities, presentation of thematic posters on soil functions, fauna and biodiversity, degradation, climate changes. At the beginning of the activities all the students were asked to fill out a questionnaire for a survey. Aim of the survey was to assess the level of knowledge of soil properties and the perception of soil functions and soil degradation.

The questionnaire, named "Do you know soil?" and integrally reported at the end of this article, consisted in 4 sections:

Section I. – "Socio-demographic characteristics", including gender, residency (town/district), type of school, parents' level of education.
Section II. – "Soil knowledge", consisting in 8 close-ended questions related to various soil properties. The questions ranged from finding the correct meaning of a soil-term to selecting the proper definition of basic soil intrinsic features. The students could tick one of four listed options.
Section III. – "Soil functions", consisting in a list of 8 statements, the agreement to which was expressed through a Likert scale ranging from 1 to 5. The statements were related to the ecological and social functions of soils.
Section IV. – "Soil perception", consisting in 5 questions aimed to get students' opinion about different topics, such as causes and consequences of soil degradation. They were asked to tick, according to their opinion, the three main causes and three main consequences of soil degradation. In the same section, they were further asked on the importance of spreading a "soil culture" within the general public. They were then invited to select, within a list, the more effective ways to pursue this objective.

A total number of about 500 high school students, aged between 16 and 18 years, participated to the survey. The type of high school ranged from scientific to agricultural, from technical to professional school.

After collecting all questionnaires, a preliminary assessment of the validity of all data was performed with the aim to eliminate those questionnaires formally incorrect, wrongly used or missing the socio-demographic information.

After this validation process the final database, including a total number of 474 observations, represented the data set used for the analyses.

All data collected were evaluated through a quantitative statistical approach. The sample characteristics were investigated through simple descriptive statistical methods while the analysis of data related to section II, III and IV was carried out through regression and correlation models using XLStat software.

2.2 Sample characteristics

The sample is homogeneous for gender (48% female, 52% male). As for residency, about one fourth of the interviewed students lives in the town of Viterbo while the rest lives in the district.

Students from the academic high school (scientific) are the most represented (53%) followed by vocational schools such as: technical (24%), professional (15%) and agricultural (8%).

Concerning parents' educational level, 15% of fathers and 17% of mothers own a university degree, while lower levels of education characterize 31% of fathers and 26% of mothers. Therefore, it can be observed that between the two parents, mothers show, on the average, a higher education level. This is not surprising, considering that this figure is quite consolidated among middle-aged Italian people, in particular those resident in the central part of the country (Istat, 2016).

Although the sample is not probabilistic, this study allows extending the results with a reasonable level of confidence for two reasons. First, the sample is not self-selected (as it happens in the internet surveys); second, the sample size is quite large, considering that the students of same age in the province of Viterbo are about 3.000. To strengthen the point of sample representativeness, it can be noticed that parents' educational level figures appear coherent with corresponding national census data (Istat, 2016).

3 RESULTS AND DISCUSSION

3.1 *Section II – soil knowledge*

The total number of correct responses to the 8 questions proposed in section II of the questionnaire accounted for less than 40% (38.7%); this result indicates a poor knowledge of soil properties and, more in general, of soil intrinsic features (Figure 1). This statement is motivated by the fact that a completely random choice within the four options proposed would have given up to 25% of correct responses.

The results shown in Figure 1 show a lack of knowledge of concepts such as pedogenesis and soil sealing while it seems less explainable the ignorance on soil depth concept, often confused with the underlying parent material. Approximately one student out of two knows that soil is a non-renewable resource. More consolidated seem to be the concepts of soil organic matter, degradation and compaction.

The students' soil knowledge level has been put in relation with the socio-demographic variables by a multiple linear regression model, where the dependent variable Y (number of correct answers) was explained by six independent variables referring to four dimensions:

– Personal characteristics – Gender (x1);
– Social environment – Residence (x2);
– Educational environment – Type of school (x3, x4);
– Cultural environment – Parents' educational level (x5, x6).

Regression model results are shown in Table 1.

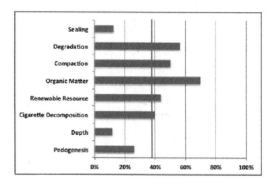

Figure 1. Section II – Soil knowledge: relative frequency of correct answers to each of the eight questions.

Table 1. Results of the regression model.

Variable	Coefficient
Intercept	+2.82**
Gender – *Male* (X_1)	−0.06
Residence – *Town* (X_2)	−0.13
School – *Scientific high school* (X_3)	+0.33**
School – *Agricultural college* (X_4)	+1.58**
Parents' education – *Father graduated* (X_5)	+0.32*
Parents' education – *Mother graduated* (X_6)	−0.11

**p-value < 0.05; *p-value < 0.1.

The corrected R-square value is quite low (0.103), a quite usual condition in social sciences studies dealing with human behaviour (Pituch & Stevens, 2016), showing that other variables, beyond those selected, influence students' soil knowledge. Conversely, the F test of the regression is very good (p-value < 0.001) indicating that the variables chosen are good predictors of soil knowledge.

Looking at the regression coefficients, the educational environment and, to a lesser extent, the family cultural level (curiously only father-related) play a significant role in increasing soil knowledge among young generations.

The role of school is certainly fundamental and acts in two directions. The first one is represented by the scientific school (belonging to the liceo category in Italy, a culturally higher level among high schools) providing a more consolidated science-oriented training. The second one relates to the agricultural school; students from this college provided 1.6 more correct answers out of 8, that is an increase by 20% of soil knowledge. This is not surprising, as the agricultural college does include elements of soil science within characterizing teaching topics. Keeping in mind the necessity to place soil information in the science-practice interface (In-gram et al., 2016), we can demonstrate that vocational schools, such as the agricultural in this case, seem more effective in promoting further training and adult competences (the so-called specialization effect) (Brunello and Chechi, 2007).

As for the significant dependence of soil knowledge from father's educational level, we may infer that men, generally, own technical-scientific university degrees (ISTAT, 2016). This result confirms that soil information, when present, is confined within specific knowledge fields that, furthermore, consider only few of its precious functions. In fact, academic degrees dealing with quantitative sciences account only for the provisioning functions of soil constituting the basis of engineering, architecture and economics.

This result provides additional evidence, and offers at the same time food for thought, that soil knowledge is still confined within a restricted vision of its multiple functions. Puig della Bellacasa, (2015) claims that the intimate relation of soil science with the predominant orientation of a techno-scientific and productionist view should be overcome to return soil its original dignity. In particular, an ethical shift of perspective is desirable. There is a need to shift from soil instrumental values—the so-called ecosystem services, blamed to represent an anthropocentric attitude (Bavaye et al., 2016) - to an ecocentric view focused on soil intrinsic value. This would imply the re-establishment of a soil ethics (Thompson, 2011), that can be attained through a social and cultural (r)evolution.

3.2 Section III – soil functions

The third section of the questionnaire aimed to assess the students' awareness of soil functions. For this purpose the students were asked to express their agreement to eight statements through a Likert scale ranging from 1 (none) to 5 (total). Figure 2 reports the relative frequency of the level of agreement to the eight items.

All statements related to soil functions reported a total (5) or high (4) level of consensus within the sample. The students showed a major perception for soil functions such as "hosts diverse forms of life" (89%) and "serves for food production" (80%). Functions such as "is responsible for water quality" and "provision of raw materials" received considerable attention, although to a lesser extent (62% and 64% respectively).

To investigate the relationship between the awareness on each soil function (expressed by the score assigned to the agreement to the related item) and the level of soil knowledge (measured by the number of correct answers to the eight questions) a correlation analysis was carried out. Significant positive Pearson correlation coefficients (p-value<0.01) were obtained for all the listed functions except for the influence of soil health on human health (data not shown). Therefore, it can be assessed that higher soil knowledge carries with it a higher awareness of its functions.

The failure to link soil properties to its influence on human health deserves specific attention. As reported by Pepper (2013), a "soil health: human health" nexus is still missing; this can be included amidst the cultural reasons, previously listed, hampering soil awareness. As an example, due to the gradual historical enfranchisement of mankind from soil, the collective imagination lost the connection between food and soil. It is therefore difficult to perceive that threats on soil health may ultimately affect human well-being (Brevik et al., 2017).

As expected, and as observed also for soil functions perception, the students from agricultural college showed a different behaviour with higher score values. It is worth to note that this happened, in particular, for functions such as "essential element of the landscape" (4.61 vs 4.09) and "source of raw material for industries" (4.12 vs 3.68).

Education is, therefore, not only the main driver for a basic knowledge of soil properties but it can help to build up a specific awareness of soil functions whenever these are strictly connected to the teaching topics characterizing the school training approach.

3.3 Section IV – soil perception

Section IV of the questionnaire consisted in various questions aimed to investigate the perception of causes and effects of soil degradation and, finally, the likely actions for reversing this process.

The students were asked to choose the three main causes of degradation within a list of seven (Figure 3).

The students identified the major causes of soil degradation in human activities that, in their imag-

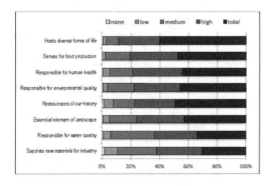

Figure 2. Section III – Soil functions. Level of agreement to eight statements related to soil functions.

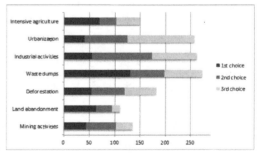

Figure 3. Section IV – Soil perception. Causes of soil degradation.

ination, are strictly linked to the general concepts of pollution and degradation of the environment. Waste dumps, industries and urbanization received, in fact, the highest scores with a total of 273, 262 and 257 preferences (as the sum of first, second and third choice). Conversely, they seem less aware that degradation may derive from the primary sector. Causes such as "deforestation", "intensive agriculture", "land abandonment" received lower scores with 182, 150 and 111, respectively. It is evident that linking land overexploitation and reduction of plant cover to soil damage is less obvious and certainly derives from the scarce knowledge of its properties.

As for the consequences of soil degradation (Figure 4), the results clearly confirm what emerged when analysing the causes. Water pollution and air pollution are markedly placed in top positions, with more than 200 preferences each, with respect to the other options proposed.

Again, the strongest perception of air and water quality issues, easily perceived within the general public (Bouma et al., 2012), drives the students choices. Furthermore, it is interesting to highlight that the students preferred consequences that affect atmosphere and hydrosphere and not the pedosphere itself. In fact, loss of structure stability, reduction of fertility and biodiversity were less considered. This can suggest that soil is perceived as a medium leading to other important Earth compartments missing, in this way, its intrinsic value.

Finally, the last part of section IV investigated over the likely actions that could help to raise soil awareness and promote soil protection. The great majority of the sample considered very important (62%) or important (27%) to inform citizens about the importance of soil as a common good and the dissemination of a soil culture (data not shown). Among the possible actions that a citizen can choose to raise soil awareness (Figure 5), the students had the possibility to select three out of five listed answers.

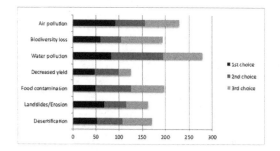

Figure 4. Section IV – Soil perception. Consequences of soil degradation.

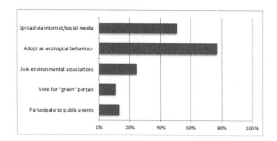

Figure 5. Section IV – Soil perception. How to promote soil protection.

An ecologically sound behaviour emerged as the most selected option. Communication by means of the net-work (mainly through social media) was indicated by about one half of the sample. The role of environmental associations seemed to be less important while political engagement and participation to public events weren't considered at all as valuable tools for soil valorisation.

As for other environmental and, more in general, social issues, it is evident that youngsters rely on individual attitudes or "virtual" (cyberspace) behaviours (Loader et al., 2014). Conversely, "real" social events or actions, such as demonstrations, associations and, mainly, political engagement, are considered out-dated and, consequently, less effective (Beck and Beck-Gernsheim, 2002).

4 CONCLUSIONS

The study reveals a weak knowledge of soil intrinsic features within high school students. However, when related to the perception of soil functions, these two aspects are positively correlated indicating that knowledge plays a crucial role in promoting soil awareness. The students' perception of soil degradation is mostly associated to other environmental issues (i.e. water and air quality), which are more consolidated concepts within the general public. Very poor is the perception of the degradation of soil intrinsic features and functions.

More than familiar background, education is the main driver and factor significantly affecting students' basic knowledge of soil properties. It is thus imperative to work in this direction because connectivity, promoted through instruction, plays a crucial and effective role to guarantee soil security.

Although this study is not intended to draw a picture of all Italian high school students' attitudes towards soil, these results provide an insight into the interrelationships of youngsters with environmental issues.

This kind of knowledge could be helpful to draw a successful approach to promote and build up a solid soil culture starting from new generations within modern society.

REFERENCES

Albayrak, A., & Hackverdi Can, M. 2012. Investigation of elementary school students' perceptions about "soil". Procedia—Social and Behavioral Sciences 46: 5635–5639

Bavaye, P.C., Bavaye, J., Gowdy, J. 2016. Soil "Ecosystem" Services and Natural Capital: Critical Appraisal of Research on Uncertain Ground. Frontiers in Environmental Science 4 (41): 1–49.

Beck, U., Beck-Gernsheim, E. (2002) Individualization: Institutionalized individualism and its social and political consequences, London, Sage.

Brunello, G. & Checchi, D. 2007. Does school tracking affect equality of opportunity? New international evidence. Ecoomic Policy 22 (52): 782–861.

Bouma, J., Broll, G., Crane, T.A, Dewitte, O., Gardi, C., Schulte R.P.O., Towers, W. 2012. Soil information in support of policy making and awareness raising. Current opinion in Environmental sustainability 4: 552–558

Brevik, Eric C., Joshua J. Steffan, Lynn C. Burgess, and Artemi Cerdà. 2017. Links Between Soil Security and the Influence of Soil on Human Health, in: Damien Field, Cristine Morgan, and Alex McBratney (Eds.), Global Soil Security. Progress in Soil Science Series, Springer, Rotterdam. p. 261–274.

Hermann, L. 2006. Soil education: a public need. Developments in Germany since the mid 1990s. Journal of Plant Nutrition and Soil Science 169: 464–471.

Istat. 2016. Italian Census Data, www.istat.it

Ingram, J., Mills, J., Dibari, C., Ferri, R., Bahadur Ghaley, B., Hansen, G., Iglesias, A., Karaczun, Z., McVittie, A., Merante, P., Molnar, A., Sánchez, B. 2016. Communicating soil carbon science to farmers: Incorporating credibility, salience and legitimacy. Journal of Rural Studies, 48: 115–128

Loader, B.D., Vromen, A., Xenos, M.A. 2014. The net-worked young citizen: social media, political participation and civic engagement. Information, Communication & Society, 17, (2): 143–150

McBratney, A. & Field, D., 2015. Securing our soil. Soil Science and Plant Nutrition, 61: 587–591.

McBratney, A., Field, D.J., Koch, A. 2014. The dimensions of soil security. Geoderma 213: 203–221

Moscatelli, M.C., De Minicis, E., Grego, S. 2011 La percezione del suolo come espressione del patrimonio sociale e culturale, In Carmelo Dazzi (Ed.) La perezione del suolo Proc. Nat. Symp., Palermo 2–3 December 2010, ISBN 978-88-95315-11-9. Le Penseur, pp. 192–196.

Pepper, I.L. 2013. The Soil Health-Human Health Nexus. Critical Reviews in Environmental Science and Technology, 2617–2652.

Pituch, K.A., Stevens J.P. 2016. Applied multivariate statistics for the social sciences. Routledge, New York.

Puig de la Bellacasa, M. 2015. Making time for soil: Technoscientific futurity and the pace of care. Social Studies of Science, 1–26.

Thompson, P.B. 2011. The Ethics of Soil. In T.J. Sauer (ed.), Sustaining Soil Production in Response to Global Climate Change: Science, Policy and Ethics. John Wiley and Sons. pp. 31–42.

Opportunities for enhancing soil health

C. Wayne Honeycutt, Steven R. Shafer, Sheldon R. Jones & Byron F. Rath
Soil Health Institute, Morrisville, North Carolina, USA

ABSTRACT: Recently, the concept of "soil health" has gained widespread interest by farmers/ranchers, scientists, service laboratories, food and apparel industries, and the general public. Such a high-level of interest is well-founded because research has shown that adopting soil health-promoting practices and management systems is not only good for the farmer, but also benefits the environment. In particular, a wealth of research shows that these soil health practices increase carbon sequestration, reduce greenhouse gas emissions, and reduce nutrient losses through both runoff and leaching. To expand on this scientific basis for soil health, the "Soil Health Institute" was created in 2015 to "Safeguard and enhance the vitality and productivity of soil through scientific research and advancement." The Institute is comprised of a diverse range of stakeholders and brings value to the soil science community and global public by identifying key gaps in research and adoption, designing strategies and obtaining/providing funds to address those gaps, and ensuring that research results are transferred into the hands of field practitioners whose effective management of our landscapes benefits us all.

1 INTRODUCTION

Soil health is defined as the continued capacity of soil to function as a vital, living ecosystem that sustains plants, animals, and humans (USDA-NRCS). The concept of soil health has gained widespread interest and acceptance by farmers and ranchers. This provides an opportunity to not only assist farmers with building yield stability and profitability, but also to address climate change adaptation/mitigation and to improve water quality through these soil health-promoting practices and systems. Numerous, peer-reviewed studies support these opportunities, as briefly summarized in the following.

2 ADDRESSING CLIMATE CHANGE ADAPTATION AND MITIGATION THROUGH SOIL HEALTH

While approximately 1% of land is generally impacted by drought, this is predicted to increase to 30% by the year 2100 given current levels of greenhouse gas emissions (Burke et al., 2006). Even for low and moderate emission levels, extreme drought is predicted to increase (Strzepek et al., 2010). For the 30-year period from 2036–2065, over 18 additional months of extreme drought is projected for major portions of the United States. For 2066–2095, the predictions are even more dire (Strzepek et al., 2010). It is important to recognize that such increases in drought will not only impact crop and livestock production, but will also negatively impact fish and wildlife habitat, water quality, air quality, wind erosion, aquifer and surface water storage, as well as civil discourse pertaining to water rights.

One of the greatest opportunities for increasing drought resilience by improving soil health resides in the relationship between soil organic carbon and a soil's capacity to hold plant-available water. An example of this relationship is provided in Fig. 1, where increasing soil organic carbon is clearly

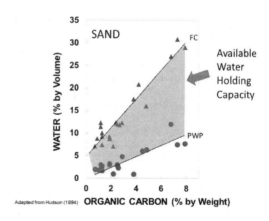

Figure 1. Relationship between soil organic C and available water holding capacity (FC = Field Capacity; PWP = Permanent Wilting Point). (Adapted from Hudson, 1994).

shown to dramatically increase available water holding capacity (Hudson, 1994).

Several studies have demonstrated this relationship. For example, research conducted by Olness and Archer (2005) showed that a 1% increase in soil organic carbon resulted in a 2 to > 5% increase in available water holding capacity. A review of several studies conducted across a range of soil textures (including sand, loamy sand, sandy loam, fine sandy loam, silt loam, loam, and silty clay), reported that a 1g increase in soil organic carbon increased available water by 1 to 4.9g for every 100g of soil (Emerson, 1995). Therefore, increasing soil organic carbon can significantly enhance resilience of our soils, cropping systems, and grazing systems to both drought and heavy precipitation.

Unfortunately, most of our cultivated soils have lost approximately 20–40% of their native organic carbon, thereby increasing crop vulnerability to extreme weather events like drought (Davidson and Ackerman, 1993). However, research conducted across a wide range of locations, soils, climates, and cropping systems has shown that soil health-promoting practices such as no-tillage, cover crops, and crop rotation can increase soil organic carbon, thereby restoring resilience to extreme weather, while simultaneously mitigating climate change through carbon sequestration.

Examples of net C sequestration through no-tillage are provided in Table 1. The preponderance of evidence clearly shows that such practices as no-tillage can increase soil organic C (SOC).

Research has also shown, however, that the rates of microbial processes involved in decomposition of crop residue, animal manure, and other organic materials (i.e. the very processes by which soil organic carbon is formed) are highly dependent on temperature, moisture, chemical composition, soil properties and other factors (Honeycutt et al., 1988; 1993; 2005; Doel et al., 1990; Griffin et al., 2002). In addition, other soil properties can influence the soil organic carbon vs. available water holding capacity relationship (Rawls, 2003; Saxton and Rawls, 2006). Consequently, additional research and education efforts are needed to provide practical tools that will allow farmers to fully capitalize on the soil organic carbon vs. available water holding capacity relationships.

Such opportunities for increasing soil organic carbon are not limited to cropland. The practices used for managing range and pasture lands not only influence the above-ground biomass, but also impact the below-ground carbon (Table 2). As demonstrated over a 21-year study, light (prescribed) grazing increased soil organic carbon even above that level maintained with no grazing at all (Table 2). Adding legumes to a pasture can also increase its overall biological productivity

Table 1. Summary of research investigating tillage impact on soil organic C (SOC).

SOC-CT Mg/ha	SOC-NT Mg/ha	Reference
23.5	26.2	Terra et al. (2005)
60.3	71.1	Karlen et al. (1994)
45.4	51.3	Mielke et al. (1986)
46.6	58.5	Yang & Wander (1999)
106.3	110.4	Wander et al. (1998)
72.5	78.6	Wander et al. (1998)
42.7	43.9	Wander et al. (1998)
60.0	73.0	Elliott et al. (1994)
77.0	65.0	Elliott et al. (1994)
47.7	46.3	Blevins et al. (1977)
45.9	52.8	Blevins et al. (1977)
61.3	66.2	Ismail et al. (1994)
48.9	55.3	Ismail et al. (1994)
53.3	62.0	Mielke et al. (1986)
30.0	37.0	Elliott et al. (1994)
40.0	37.0	Elliott et al. (1994)
63.3	75.3	Varvel & Wilhelm (2010)
58.7	66.2	Varvel & Wilhelm (2010)
61.5	72.7	Varvel & Wilhelm (2010)
45.3	80.0	Ussiri & Lal (2009)
76.8	117.7	Dendooven et al. (2012)
13.8	15.9	Deen & Kataki (2003)
49.4	58.3	Malhi & Kutcher (2007)
75.2	79.6	Malhi & Kutcher (2007)
22.2	26.6	Hernanz et al. (2002)
19.4	25.6	Hernanz et al. (2002)
46.0	49.9	Morell et al. (2011)

CT = Conventional Tillage; NT = No-Tillage.

Table 2. Impact of grazing intensity on soil organic carbon and soil nitrogen. (Adapted from Ganjegunte et al., 2005).

Treatment	Soil organic C, 0–5 cm (Mg/ha)*	Soil N, 0–5 cm (Mg/ha)
No grazing	10.8 b	0.94 b
Light grazing	13.8 a	1.23 a
Heavy grazing	10.9 b	0.94 b

*Numbers within a column followed by the same lower case letter are not significantly different at the 0.05 level.

(Mortensen et al., 2004), thereby increasing a soil's organic carbon content and imparting resilience to extreme drought and heavy precipitation.

In addition to increasing available water holding capacity, these soil-health promoting practices, such as no-till, cover crops, and crop rotation, can also significantly increase water infiltration. Research studies have documented no-till in continuous corn to increase infiltration rate by 165%

and no-till in a wheat-sorghum-fallow system to increase infiltration rate by 132–194% (TerAvest et al., 2015; Stone and Schlegel, 2010). Adding a cover crop to a no-till system brings additional benefits, with increased infiltration rate ranging from 164–462% (Blanco-Canqui et al., 2011; Steele et al., 2012).

Increased infiltration observed with no-tillage and cover crops is also reflected in reduced runoff. For example, studies in Georgia found 2.6 times more runoff with conventional tillage than with no-tillage (Endale et al., 2010). Reduced runoff and increased infiltration provided by no-till have not only been shown on research plots, but also at the watershed scale. On 10–15% slopes in Ohio, Edwards et al. (1988) reported that a 50-year storm produced 15 times more runoff in a conventionally tilled watershed than a no-till watershed. Over 46% of the rainfall ran off the conventionally tilled watershed, while less than 3% of the rainfall ran off the no-till watershed. Consequently, this greater infiltration not only improves resilience to extreme weather by recharging soil water and underlying aquifers, but the reduced runoff also reduces soil and nutrient losses to surface water (Yoo et al., 1988; Zhu et al., 1989; Sharpley and Smith, 1991). Such water quality benefits derived through soil health are described below.

3 ADDRESSING WATER QUALITY THROUGH SOIL HEALTH

Cover crops are a valuable tool for not only enhancing soil health, but for also reducing nutrient losses through surface runoff and leaching. As demonstrated across a wide range of soils and climates, peer-reviewed scientific research has repeatedly shown cover crops to reduce nitrate leaching losses by over 60% (Table 3). Indeed, a meta-analysis of 69 separate studies from across the U.S. showed that cover crops reduced nitrate leaching losses by an average of 70% (Tonitto et al., 2006).

Not only have these benefits been demonstrated across a wide range of soils and climates, but they have also been shown to persist from year to year. An example is provided from a research study conducted in Iowa, where although variable weather from one year to the next resulted in different amounts of nitrate leaching, cover crops consistently reduced those levels in each year of the study (Table 4).

Additional soil health practices, such as crop rotation and nutrient management, also play key roles for protecting and enhancing water quality through soil health. In fact, reduction in nitrate loss to groundwater is generally due to plant uptake of that nitrate by the cover crop (Jewett

Table 3. Summary of literature demonstrating cover crop impacts on nitrate leaching (Adapted from Kaspar and Singer, 2011).

Location	Cover crop	Reduction in Nitrate leaching (%)	Reference
California	Rye	65–70	Wyland et al. (1996)
Delaware	Rye	30	Ritter et al. (1998)
France	Ryegrass	63	Martinez and Guirard (1990)
Indiana	Winter Wheat (and reduced fertilizer)	61	Kladivko et al. (2004)
Iowa	Rye	61	Kaspar et al. (2007)
Kentucky	Rye	94	McCracken et al. (1994)
Kentucky	Hairy Vetch	48	McCracken et al. (1994)
Maryland	Rye	77	Staver and Brinsfield (1990)
Maryland	Rye	80	Staver and Brinsfield (1998)
Michigan	Rye	28–68	Rasse et al. (2000)
Minnesota	Rye	13	Strock et al. (2004)

Table 4. Effects of cover crops on nitrate levels exiting subsurface drainage tile in Iowa, 2002–2005 (Adapted from Kaspar et al., 2007).

Cumulative nitrate load (kg N ha^{-1})

Year	No cover crop	Gamagrass cover crop	Rye cover crop
2002	40.4a*	36.3a	11.2b
2003	81.1a	57.3a	33.9b
2004	47.2a	37.8a	23.0b
2005	34.4a	19.6ab	11.1b
Average	50.8a	37.7a	19.8b

*Numbers within a row followed by the same lower case letter are not significantly different at the 0.05 level.

and Thelen, 2006). Consequently, in order to ensure benefits to water quality, fertilizer recommendations for the following crop must account for nitrogen that will become available from the decomposing cover crop. As many practices used to enhance soil health influence nutrient availability, nutrient management is inherently a key soil health practice, and an entire systems perspective is required to optimize benefits and minimize nutrient losses. This recognition has given rise to the "4R" nutrient stewardship concept advocated by the fertilizer industry and followed by many natural resource professionals and farmers (Mattson and van Iersel, 2011), where nutrients are to be added from the Right source, at the Right rate and Right time, and in the Right place (4R Nutrient Stewardship).

4 THE SOIL HEALTH INSTITUTE

To build and expand on this scientific basis, the Soil Health Institute was created in 2015 to *"Safeguard and enhance the vitality and productivity of soil through scientific research and advancement."* As the independent, non-profit organization charged with coordinating and supporting soil stewardship and advancing soil health, the Soil Health Institute (SHI) is focused on fundamental, translational, and applied research and ensuring its adoption. Its premise is that soil health must emerge as the cornerstone of land use management decisions throughout the world during the 21st century because healthy soil is the foundation of life and society. Enhancing soil health allows us to improve water quality, increase drought resilience, reduce greenhouse gas emissions, improve farm economies, provide pollinator habitat, and better positions us to feed the nine billion people expected by 2050.

To operationalize such opportunities, the Soil Health Institute works with a broad base of stakeholders to:

1. Identify gaps in research and adoption;
2. Build research/implementation strategies and their corresponding networks necessary to appropriately address those gaps;
3. Seek and obtain funding to address the above gaps;
4. Administer an accountable, transparent, and technically proficient competitive grants program;
5. Ensure impact of the investments;
6. Incorporate research results into educational materials; and
7. Enhance partnerships for increasing technology transfer and adoption.

Accordingly, the Soil Health Institute has incorporated input from numerous stakeholders and developed *"Enriching Soil, Enhancing Life: An Action Plan for Soil Health",* which is available at http://soilhealthinstitute.org/. The Action Plan not only identifies gaps in key areas of research, measurements, economics, communication/education, and policy; but also lists specific actionable steps needed to address those gaps.

Collectively, these efforts are designed to move knowledge and technology from the research laboratory to the farm field by bringing together traditional and non-traditional agricultural industry partners, farmers, ranchers, government agencies, scientists, and consumers to focus on one common, clear goal: protecting and enriching our soils, our environment, and indeed, our lives. All who wish to participate in this journey are invited to join and help make soil health the cornerstone for managing our natural resources around the world. For more information, please visit http://soilhealthinstitute.org/.

REFERENCES

4R Nutrient Stewardship. http://www.nutrientstewardship.com/what-are-4rs.

Baumhardt, R.L., G.L. Johnson, and R.C. Schwartz. 2011. Residue and long-term tillage and crop rotation effects on simulated rain infiltration and sediment transport. *Soil Sci. Soc. Am. J.* 76:1370–1378.

Blanco-Canqui, H., M.M. Mikha, D.R. Presley, and M.M. Claassen. 2011. Addition of cover crops enhances no-till potential for improving soil physical properties. *Soil Sci. Soc. Am. J.* 75:1471–1482.

Blevins, R.L., G.W. Thomas, and P.L. Cornelius. 1977. Influence of no-tillage and nitrogen fertilization on certain soil properties after 5 years of continuous corn. *Agron. J.* 69:383–386.

Dabney, S.M., J.A. Delgado, and D.W. Reeves. 2001. Using winter cover crops to improve soil and water quality. *Commun. Soil Sci. Plant Anal.* 32:1221–1250.

Davidson, E.A., and I.L. Ackerman. 1993. Changes in soil carbon inventories following cultivation of previously untilled soils. *Biogeochem.* 20:161–193.

Deen, W., and P.K. Kataki. 2003. Carbon sequestration in a long-term conventional versus conservation tillage experiment. *Soil Tillage Res.* 74:143–150.

Dendooven, L., L. Patino-Zuniga, N. Verhulst, M. Luna-Guido, R. Marsch, and B. Govaerts. 2012. Global warming potential of agricultural systems with contrasting tillage and residue management in the central highland of Mexico. *Agric. Ecosyst. Environ.* 152:50–58.

Doel, D.S., C.W. Honeycutt, and W.A. Halteman. 1990. Soil water effects on the use of heat units to predict crop residue carbon and nitrogen mineralization. *Biol. Fertil. Soils* 10:102–106.

Edwards, W.M., M.J. Shipitalo, and L.D. Norton. 1988. Contribution of macroporosity to infiltration into a continuous corn no-tilled watershed: implications for contaminant movement. *J. Contam. Hydrol.* 3:193–205.

Elliott, E.T., I.C. Burke, C.A. Monz, S.D. Frey, D.J. Lyon, K. Paustian, H.P. Collins, A.D. Halvorson, D.R. Huggins, E.A. Paul, R.F. Turco, C.V. Cole, M.V. Hickman, R.L. Blevins, and W.W. Frye. 1994. Terrestrial carbon pools: Preliminary data from Corn Belt and Great Plains regions. pp. 179–191. In: *Defining soil quality for a sustainable environment. Soil Sci. Soc. Am. Spec. Pub. No. 35.* SSSA Madison, WI.

Emerson, W.W. 1995. Water retention, organic C and soil texture. *Aust. J. Soil Res.* 33:241–251.

Endale, D.M., H.H. Schomberg, M.B. Jenkins, D.H. Franklin, and D.S. Fisher. 2010. Management implications of conservation tillage and poultry litter use for Southern Piedmont USA cropping systems. *Nutr. Cycl. Agroecosyst.* 88: 299–313.

Ganjegunte, G.K., G.F. Vance, C.M. Preston, G.E. Schuman, L.J. Ingram, P.D. Stahl, and J.M. Welker. 2005. Soil organic carbon composition in a northern mixed-grass prairie: Effects of grazing. *Soil Sci. Soc. Am. J.* 69:1746–1756.

Griffin, T.S., C.W. Honeycutt, and Z. He. 2002. Effects of temperature, soil water status, and soil type on swine slurry nitrogen transformations. *Biol. Fertil. Soils* 36:442–446.

Hernanz, J.L., R. Lopez, L. Navarrete, and V. Sanchez-Giron. 2002. Long-term effects of tillage systems and rotations on soil structural stability and organic carbon stratification in semiarid central Spain. *Soil Tillage Res.* 66:129–141.

Honeycutt, C.W., T.S. Griffin, and Z. He. 2005. Manure nitrogen availability: Dairy manure in northeast and central U.S. soils. *Biol. Agric. Hort.* 23:199–214.

Honeycutt, C.W., L.J. Potaro, K.L. Avila, and W.A. Halteman. 1993. Residue quality, loading rate and soil temperature relations with hairy vetch (*Vicia villosa* Roth) residue carbon, nitrogen and phosphorus mineralization. *Biol. Agric. Hort.* 9:181–199.

Honeycutt, C.W., L.M. Zibilske, and W.M. Clapham. 1988. Heat units for describing carbon mineralization and predicting net nitrogen mineralization. *Soil Sci. Soc. Am. J.* 52:1346–1350.

Hudson, B.D. 1994. Soil organic matter and available water capacity. *J. Soil Water Cons.* 49:189–194.

Ismail, I., R.L. Blevins, and W.W. Frye. 1994. Long-term no-tillage effects on soil properties and continuous corn. *Soil Sci. Soc. Am. J.* 58:193–198.

Jewett, M.R., and K.D. Thelen. 2007. Winter cereal cover crop removal strategy affects spring soil nitrate levels. *J. Sustainable Agric.* 29:55–67, doi: 10.1300/J064v29n03_06.

Karlen, D.L., N.C. Wollenhaupt, D.C. Erbach, E.C. Barry, J.B. Swan, N.S. Nash, and J.L. Jordan. 1994. Long-term tillage effects on soil quality. *Soil Tillage Res.* 32:227–313.

Kaspar, T.C., D.B. Jaynes, T.B. Parkin, and T.B. Moorman. 2007. Rye cover crop and gamagrass strip effects on NO_3 concentration and load in tile drainage. *J. Environ. Qual.* 36:1503–1511.

Kaspar, T.C., and J.W. Singer. 2011. The use of cover crops to manage soil. http://digitalcommons.unl.edu/cgi/viewcontent.cgi?article=2387&context=usdaarsfacpub.

Kladivko, E.J., J.R. Frankenberger, D.B. Jaynes, D.W. Meek, B.J. Jenkinson, and N.R. Fausey. 2004. Nitrate leaching to subsurface drains as affected by drain spacing and changes in crop production system. *J. Environ. Qual.* 33:1803–1813.

Kunkel, K.E., T.R. Karl, H. Brooks, J. Kossin, J.H. Lawrimore, D. Arndt, L. Bosart, D. Changnon, S.L. Cutter, N. Doesken, K. Emanuel, P. Ya. Groisman, R.W. Katz, T. Kunitson, J. O'Brien, C.J. Paciorek, T.C. Peterson, K. Redmond, D. Robinson, J. Trapp, R. Vose, S. Weaver, M. Wehner, K. Wolter, and D. Wuebbles. 2013. Monitoring and understanding trends in extreme storms: State of knowledge. *Am. Meteor. Soc.* doi: 10.1175/BAMS-D-11-00262.1.

McCracken, D.V., M.S. Smith, J.H. Grove, R.L. Blevins, and C.T. MacKown. 1994. Nitrate leaching as influenced by cover cropping and nitrogen source. *Soil Sci. Soc. Am. J.* 58:1476–1483.

Malhi, S.S., and H.R. Kutcher. 2007. Small grains stubble burning and tillage effects on soil organic C and N, and aggregation in northeastern Saskatchewan. *Soil Tillage Res.* 94:353–361.

Martinez, J., and G. Guiraud. 1990. A lysimeter study of the effects of a ryegrass catch crop, during a winter-wheat maize rotation, on nitrate leaching and on the following crop. *J. Soil Sci.* 41:5–16.

Mattson, N.S., and M.W. van Iersel. 2011. Application of the "4R" nutrient stewardship concept to horticultural crops: applying nutrients at the "right time". *HortTechnology* 21:667–673.

Mielke, L.K., J.W. Doran, and K.A. Richards. 1986. Physical environment near the surface of plowed and no-tilled soils. *Soil Tillage Res.* 7:355–366.

Morell, F.J., C. Cantero-Martinez, J. Lampurlanes, D. Plaza-Bonilla, and J. Alvaro-Fuentes. 2011. Soil carbon dioxide flux and organic carbon content: Effects of tillage and nitrogen fertilization. *Soil Sci. Soc. Am. J.* 75:1874–1884.

Mortenson, M.C., G.E. Schuman, and L.J. Ingram. 2004. Carbon sequestration in rangelands interseeded with yellow-flowering alfalfa (*Medicago sativa* ssp. *falcata*). *Environ. Manage.* 33:S475–S481.

Olness, A., and D. Archer. 2005. Effect of organic carbon on available water in soil. *Soil Sci.* 170:90–101.

Rasse, D.P., J.T. Ritchie, W.R. Peterson, J. Wei, and A.J.M. Smucker. 2000. Rye cover crop and nitrogen fertilization effects on nitrate leaching in inbred maize fields. *J. Environ. Qual.* 29:298–304.

Rawls, W.J., Y.A. Pachepsky, J.D. Ritchie, T.M. Sobecki, and H. Bloodworth. 2003. Effect of soil organic carbon on soil water retention. *Geoderma* 116:61–76.

Ritter, W.F., R.W. Scarborough, and A.E.M. Chirnside. 1998. Winter cover crops as a best management practice for reducing nitrogen leaching. *J. Contam. Hydrol.* 34:1–15.

Saxton, K.E., and W.J. Rawls. 2006. Soil water characteristic estimates by texture and organic matter from hydrologic solutions. *Soil Sci. Soc. Am. J.* 70:1569–1578.

Sharpley, A.N., and S.J. Smith. 1991. Effects of cover crops on surface water quality. pp.41–49 In W.L. Hargrove (ed.) *Cover crops for clean water.* Soil and Water Conservation Society, Ankeny, IA.

Staver, K.W., and R.B. Brinsfield. 1990. Patterns of soil nitrate availability in corn production systems: implications for reducing groundwater contamination. *J. Soil Water Conserv.* 45:318–323.

Staver, K.W., and R.B. Brinsfield. 1998. Using cereal grain winter cover crops to reduce groundwater nitrate contamination in the mid-Atlantic coastal plain. *J. Soil Water Conserv.* 53:230–240.

Steele, M.K., F.J. Coale, and R.L. Hill. 2012. Winter annual cover crop impacts on no-till soil physical properties and organic matter. *Soil Sci. Soc. Am. J.* 76:2164–2173.

Stone, L.R., and A.J. Schlegel. 2010. Tillage and crop rotation phase effects on soil physical properties in the West-Central Great Plains. *Agron. J.* 102:483–491.

Strock, J.S., P.M. Porter, and M.P. Russelle. 2004. Cover cropping to reduce nitrate loss through subsurface drainage in the northern U.S. Corn Belt. *J. Environ. Qual.* 33:1010–1016.

Strzepek, K., G. Yohe, J. Neumann, and B. Boehlert. 2010. Characterizing change in drought risk for the United States from climate change. *Environ. Res. Lett.* 5:1–9. doi: 10.1088/1748-9326/5/4/044012.

TerAvest, D., L. Carpenter-Boggs, C. Thierfelder, and J.P. Reganold. 2015. Crop production and soil water management in conservation agriculture, no-till, and conventional tillage systems in Malawi. *Agric. Ecosyst. Environ.* 212:285–296.

Terra, J.A., D.W. Reeves, J.N. Shaw, and R.J. Raper. 2005. Impacts of landscape attributes on carbon sequestration during the transition from conventional to conservation management practices on a Coastal Plain field. *J. Soil Water Conserv.* 60:438–446.

Tonitto, C., M.B. David, L.E. Drinkwater. 2006. Replacing bare fallows with cover crops in fertilizer-intensive cropping systems: a meta-analysis of crop yield and N dynamics. *Agric. Ecosyst. Environ.* 112:58–72.

USDA-NRCS. https://www.nrcs.usda.gov/wps/portal/nrcs/main/soils/health/.

USDA-SARE and Conservation Technology Information Center. 2013. *2012–2013 Cover crop survey*. http://www.sare.org/Learning-Center/From-the-Field/North-Central-SARE-From-the-Field/2012-Cover-Crop-Survey-Analysis.

Ussiri, D.A.N., and R. Lal. 2009. Long-term tillage effects on soil carbon storage and carbon dioxide emissions in continuous corn cropping systems from an alfisol in Ohio. *Soil Tillage Res.* 104:39–47.

Varvel, G.E., and W.W. Wilhelm. 2010. Long-term soil organic carbon as affected by tillage and cropping systems. *Soil Sci. Soc. Am. J.* 74:915–921.

Wander, M.M., M.G. Bidar, and S. Aref. 1998. Tillage impacts on the depth distribution of total and particulate organic matter in three Illinois soils. *Soil Sci. Am. J.* 62:1704–1711.

Wyland, L.J., L.E. Jackson, W.E. Chaney, K. Klonsky, S.T. Koike, and B. Kimple. 1996. Winter cover crops in a vegetable cropping system: Impacts on nitrate leaching, soil water, crop yield, pests and management costs. *Agric. Ecosyst. Environ.* 59:1–17.

Yang, X.M., and M.M. Wander. 1999. Tillage effects on soil organic carbon distribution and storage in a silt loam soil in Illinois. *Soil Tillage Res.* 52:1–9.

Yoo, K.H., J.T. Touchton, and R.H. Walker. 1988. Runoff, sediment and nutrient losses from various tillage systems of cotton. *Soil Tillage Res.* 12:13–24.

Zhu, J.C., C.J. Gantzer, S.H. Anderson, E.E. Alberts, and P.R. Beuselinck. 1989. Runoff, soil, and dissolved nutrient losses from no-till soybean with winter cover crops. *Soil Sci. Soc. Am. J.* 53:1210–1214.

The soil certificate—a Flemish tool helping raise awareness about soil

J. Ceenaeme, G. Van Gestel & N. Bal
OVAM – The Public Waste Agency of Flanders, Mechelen, Flanders, Belgium

W. Van Den Driessche
The Association of Accredited Soil Remediation Experts (VEB), Antwerp, Flanders, Belgium

ABSTRACT: Since 1995, a soil certificate must accompany each transfer of land in Flanders, with information on soil contamination. The soil certificate is a legally obliged information tool to protect acquirers that helps to raise awareness: 1) of the possible impact of soil contamination on the 'economic' value of land. It gives a stimulus to owners to take good care of the soil in order to enjoy the benefits of their land, within the framework of the free market principles; 2) of the importance of soil quality for a healthy living environment.

The content of the soil certificate changes continuously. By highlighting land as a common and adding information to the certificate that raises awareness of these issues, OVAM stimulates the owner to consider not only his 'property rights' but also his 'property obligations' towards society, to appreciate the value of the soil and land, and take good care of it.

1 THE FLEMISH SOIL POLICY AND THE PUBLIC WASTE AGENCY OF FLANDERS (OVAM) IN ITS LARGER POLITICAL CONTEXT

1.1 Flanders within Europe

Belgium has a complex political structure: what began in 1830 as a unitarian monarchist state, changed over the next 2 centuries to a federal state consisting of 3 'Regions' (the Flemish, the Walloon and the Brussels-Capital Region) and 3 'Communities' (the Flemish, the French and the German-speaking Community), each with its own parliament and government. While the division into communities resulted from a demand for more cultural autonomy, the emergence of the regions was mainly a result of the claim for more economic autonomy. Due to a significant change in the state structure (dating from the 1980 Reform of the State), the authorities of the Belgian unitarian state were divided between the federal state, the regions and the communities. In this way, the policy area 'Environment and Nature' (including the waste and soil regulations) became a strictly regional issue. As a result, Flanders (like Wallonia and Brussels) could from then on chart its own course on soil policy.

1.2 The public waste agency of Flanders and the creation of a Flemish soil policy

One of the major challenges confronting the new Flemish Government in the early 1980s, was the rapid increase in waste production and the related increase in uncontrolled landfills and dumpsites.

In order to cope with this new societal challenge, the 'Decree of July 2nd 1981 on the prevention and management of waste' (brief: 'Waste Decree') was enacted in 1981, establishing a single 'waste society' for the entire Flemish territory: the 'Public Waste Society for the Flemish Region'. OVAM was born.

Initially, OVAM's task was mainly waste-oriented: waste prevention, stimulating the reuse and recycling of waste and, if necessary, the removal of dumpsites that posed a threat to public health or to the environment.

With the latter task, it became clear that in many cases, pollution had often spread to adjacent areas. OVAM quickly acknowledged the causality between the contaminants present in the landfills and the pollution found. On the basis of an article from the Waste Decree that states that OVAM is obliged to remove all contamination that poses a threat to public health or to the environment, OVAM took its responsibility and increasingly paid attention to soil contamination. At the same time, in Flanders (as elsewhere in Europe) the negative effects of decades of industrial activity on soil quality became more and more visible, requiring the authorities to react appropriately and demanding action from the involved parties. Putting this into practice was easier said than done, however. This confronted OVAM more than once with the limits of the existing legislation. As such, it became clear that a specific soil policy and a separate soil regulation had become indispensable.

2 THE FLEMISH SOIL POLICY: FROM 'WASTE DECREE' TO 'SOIL REMEDIATION DECREE' TO 'SOIL DECREE'

2.1 Goals

This specific soil regulation came with the 'Decree of February 22nd 1995 on soil remediation' (brief: 'Soil Remediation Decree').

The Soil Remediation Decree had two main objectives: soil remediation and the prevention of soil pollution on the one hand, and the protection of new landowners against the (unexpected) presence of soil contamination on the other. Because the decree-maker was aware of the radical break with the past—any newly created contamination had to be removed immediately and completely—a less stringent approach for contamination that had originated in the past was opted for. For this reason, the terms 'new' and 'historical' contamination were introduced and the enactment date of the Soil Decree (29 October 1995) determined the qualification. In addition, the government's ambition was to have at least started the remediation of all historical contaminations by 2036.

In order to meet these goals and this ambition, an inventory of all contaminated sites was required. For this purpose, there had to be a way to examine the so-called 'risk entailing sites' (sites on which potential soil-polluting activities took place or had taken place in the past) – after all, one of the major problems that had led to the Soil Remediation Decree was the inability to force a party to tackle soil contamination or to have a site examined. To achieve this, the Soil Remediation Decree defined the concept of a 'soil examination obligation', entailing a mandatory "exploratory soil examination" by the transferor of the risk-entailing land. If soil contamination was found, a 'soil remediation obligation' was established, meaning the transfer could not take place before the other decretal obligations were met. In practice, this soil remediation obligation was imposed on the transferor of the land but the Soil Remediation Decree (and later the Soil Decree) provided a number of grounds for exemption. The party to whom this obligation was assigned retained at all times the right to recover the costs from the party that had caused the soil pollution (on the basis of the Belgian Civil Code).

The Soil Remediation Decree remained (more or less) unchanged for about a decade, but around 2005 it became clear that it was no longer in touch with the rapidly changing economic and social reality. The Soil Remediation Decree was thoroughly revised and, in 2008, it was replaced by the 'Decree of October 27th 2006 on Soil Remediation and Soil Protection' (brief: 'Soil Decree').

The main guidelines of the Soil Remediation Decree were retained, but the new Soil Decree provided a simplification of existing procedures, less stringent and/or risk-based remediation targets and a number of additional instruments to make the financial burdens more bearable. The 1995 Soil Remediation Decree had, in particular, an impact on companies in operation and on land transfers; abandoned and/or unused sites -as a result of a severe soil contamination- mainly stayed out of sight. The Soil Decree tried to provide a solution for this problem. In addition, the Soil Decree incorporated a newly added chapter on soil protection, shifting the focus from soil remediation to an integrated approach on soil management, paying particular attention to prevention.

2.2 Instruments

2.2.1 The 'transfer of land'

As mentioned above, the 'transfer of land' was the crucial policy lever to achieve the defined goals. In order to fully understand this instrument, it is important to know that the Soil Decree and the Executive Decisions explicitly define what is considered as 'land' and which legal acts are considered as 'land transfer'. Its provisions apply only when there is a 'transfer of land' in the legal sense of the Soil Decree. In all other cases, the Soil Decree is not applicable. A consequence of this distinction in legal acts regarded as 'transfer of land' is that the party considered as 'transferor', is also situation-dependent. For example, in the case of the sale of land, the owner is the transferor, but when a tenancy ends, the tenant is designated as the transferor. The latter is important as it is on the transferor's initiative -and at his or her expense- that an exploratory soil examination will be conducted. In the case of a land transfer (as stipulated by the Decree), a distinction is made between the transfer of 'risk-entailing land' and 'non-risk-entailing land', with risk-entailing land being defined in the Decree as 'land on which risk-entailing facilities are or were present'. The Decree also defines what is meant by 'risk-entailing facilities': factories, workshops, warehouses, machinery, industrial installations and operations that carry a higher risk of soil contamination and that appear on the list drawn up by the Flemish Government.

When transferring non-risk-entailing land, the transferor has only one obligation: the notary (or the real estate agent) charged with the sale must deliver a recent soil certificate (see below) to the potential acquirer before signing the agreement. The content of the soil certificate must also be included in the property sale deed. When transferring risk-entailing land, in addition to the obligation to submit a recent soil certificate, the report on a recently conducted exploratory soil examination must be present.

If this exploratory soil examination reveals the presence of a soil contamination, the transfer cannot take place, except when the transferor can demonstrate that one of the exemptions applies. In that case, the Soil Remediation Obligation shifts to another party (e.g. a user of the land in question). If the soil remediation obligation is imposed on the transferor, the transfer can only take place when all other obligations imposed by the Soil Decree are met. In case of an urgent transfer, the transferor can make use of the so-called 'accelerated transfer procedure'.

The moment of the transfer of land was chosen as a policy lever for a twofold reason: at the time of transfer, significant financial resources become available (which, if necessary, can be used by the transferor to finance soil remediation), and the sheer frequency of transfers enabled OVAM to accelerate and continuously update its inventory process. By stipulating that all decretal obligations had to be fulfilled prior to the transfer, OVAM prevented contaminated land from being transferred to insolvent acquirers (which would result in community costs). On the other hand, potential acquirers were given the opportunity to, estimate possible future costs prior to the acquisition.

In the above-mentioned procedure, the importance of the notary should not be underestimated: as a notary is the only functionary authorized to execute sales deeds, he or she has to decide whether the transferred site is risk-entailing land or not and, consequently, whether the transfer can take place or not.

2.2.2 *The soil certificate*
2.2.2.1 Goal and evolution

In the transfer procedure, an important role is played by the soil certificate: for a legally valid transfer of any type of 'land' (risk-entailing or not), a recent soil certificate must be present at the time of the execution of the deed. This obligation goes back to one of the original basic objectives of the soil legislation, namely to protect the acquirer by providing useful and objective information on the soil quality of the land concerned. With this information, the potential acquirer is able to estimate any subsequent costs related to the acquisition of the land, which provides a stronger negotiating position. In this way, a wider soil awareness is created: the real estate market becomes aware of the value-influencing effect of soil contamination, and property owners become increasingly aware of the benefits of good soil care. In other words: the impact of soil contamination on property value has become important to such an extent that free market forces—even more than the obligations imposed by the government—encourage current and future land owners to prevent their soil from becoming contaminated.

In order to draw up the inventory of all contaminated sites in Flanders, the Soil Remediation Decree provided the so-called 'Register of Contaminated Lands (abbreviated: RoCL). This Register contained information—which information was specified in the Soil Remediation Decree- on all sites where pollution exceeding certain threshold values was found; thresholds that did not necessarily imply a need for remediation. The RoCL was managed by OVAM and the data included came from all of the research reports submitted to OVAM. The data on a particular site could then be retrieved by means of the soil certificate. In other words, the soil certificate became the physical carrier that enabled the exchange of data from the RoCL. Furthermore it was (and still is) the only official, printed, document that bundles all currently known information related to the soil quality of a plot of land.

In 1995, the Soil Remediation Decree introduced the concept of 'soil certificate' and stated which information it had to contain. Although this was not explicitly stated in the Soil Remediation Decree, OVAM published two types of certificates: certificates containing information on 'filed' sites (i.e.: sites that have been examined) and so-called 'blank' certificates when no information concerning the soil quality is available yet.

When the Soil Decree came into force in 2008, there was initially only limited change in the soil certificate. The RoCL was replaced by the 'Land Information Register' (LIR) because, from that moment on, all sites that OVAM received data on (and thus not only the 'contaminated' ones) were filed in the OVAM-database.

In 2016 the soil certificate did change significantly. The reason for this was an OVAM policy decision to prioritize the development of the Municipal Inventories. Municipalities were already obliged from 1998 on to create a 'Municipal Inventory' ('MI' – an inventory of all risk-entailing land within the municipal boundaries), but most of them had failed to fulfill this obligation. By supporting the municipalities, OVAM speeded up the process and, from June 2016 on, soil certificates contain information originating from the MI. This new information made it easier for notaries to decide whether an exploratory soil examination was mandatory or not.

Nowadays, OVAM still delivers file-bound and blank soil certificates. All certificates are either delivered ex-officio (e.g. when land is first filed in the LIR), or upon request and against payment (e.g. in the context of a land transfer). Upon special request, it is possible to acquire a soil certificate for only part of a land parcel.

The current blank soil certificate contains information on the geographical location, the cadastral information, and it mentions that there is no other information on that land available at OVAM. The file-bound soil certificate states, in addition to the geographical location and the cadastral information, all information obtained from the MI and whether any contamination is present that demands further action. Additionally, all research reports approved by OVAM are listed.

2.2.2.2 The future of the soil certificate

The Soil Decree made risk-based remediation legally possible by redefining the Flemish general remediation targets. From that moment on, contamination should only be remediated to reach a level that no longer poses a threat to human health or to the environment. The risks are related to and determined by the current and future use of a site. The aim of this shift to risk-based remediation is to make soil remediation economically more efficient and, consequently, to increase the number of remediation works by making the process financially more feasible. A possible side effect, however, is that a certain residual contamination remains present. This residual contamination requires no further action and, until recently, no information about it was mentioned in the soil certificate. Thus, acquirers and real estate developers were often faced with problems such as unexpectedly high costs for the disposal and/or cleaning of contaminated soil. For instance, a result of changes in spatial planning can be that the applied risk-thresholds are no longer valid, which might lead to new risks and thus extra costs for additional remediation. To avoid this kind of situation, OVAM decided to provide additional information on the soil certificate concerning soil and groundwater pollution. More in particular, specific information on the possible impact of (residual) pollution on the use of the site is added. For example: what are the consequences of excavations, of groundwater extraction,...

In the meantime, OVAM has formulated a set of clear and uniform usage recommendations. These usage recommendations are intended to make the acquirer aware of possible future (negative) side-effects (such as remediation costs, decretal duties, limitations in use,...) resulting from the presence of residual contamination. For example, a usage recommendation for a residential area may imply that a new risk-assessment should be carried out if an existing surface covering is removed. OVAM is now examining how it can add these usage recommendations to the soil certificate in an automated way, without decreasing its legibility and accuracy.

OVAM has the tradition of regularly consulting with various stakeholders (acquirers, soil remediation experts and contractors), asking for formal or informal feedback on its policy instruments. This was also the case with the soil certificate, which resulted in adjustments related to the clarity and comprehensibility of the information provided. Technological evolutions allow more (and more complex) data to be filed and listed in the soil certificate, but that also increases the challenge of bringing the information to the user in a clear and comprehensible way. Customer inquiries and (self-)evaluation will therefore remain essential.

2.2.2.3 Intermezzo: how can the Flemish soil certificate help to reach the goals of the soil security concept and vice versa?

Soil security is an overarching concept, concerned with the maintenance and improvement of the global soil resource to produce food, fiber and fresh water, contribute to energy and climate sustainability, and to maintain the biodiversity and the overall protection of the ecosystem. The concept of soil security acknowledges the five dimensions of capability, condition, capital, connectivity and codification, of soil entities which encompass the social, economic and biophysical sciences and recognize policy and legal frameworks (McBratney et al. 2014).

Flanders is a highly industrialized and densely populated region. Raising awareness amongst stakeholders (industry, real estate, citizens, ...) on the importance of soil quality (i.e. the value of a healthy soil) and integrating this into policy and legislation is crucial for mobilizing all stakeholders to take care of the soil. This approach resonates with and corresponds to the soil security dimensions of codification and connectivity: 1) as a codification tool, the soil certificate is a legal obligation when a parcel is sold. The certificate gives objective information on the soil condition aiming to inform both the transferor and acquirer and to protect the acquirer; 2) in order to improve the lack of recognition of soil services and soil goods by the property owners and society, the soil certificate contributes also to connectivity and wants to help raise awareness on the impact of soil contamination by giving clear and objective advice on the possible risks, and it can serve as a manual on how to use your soil.

3 CONCLUSIONS

Since 1996, OVAM has delivered more than 4 million soil certificates, which amounts to around 200,000 blank and 17,000 file-bound certificates each year. The soil certificate evolved from a mainly informative standard document to an awareness-raising certificate, tailor-made for any specific user. The financial revenue from the retributions is used by OVAM to optimize its data management with maximum profits for society and to financially support non-liable owners with their remediation works.

Due to the soil certificate, OVAM succeeded, as a governmental agency, in making Flemish citizens and entrepreneurs aware of the soil legislation and the impact of soil contamination (and associated liabilities) on the financial and economic value of their land. The certificate stimulates landowners to take care of their soil, and this within the principles of the free market. The better you take care of the soil, the more you can enjoy its benefits. Objective information also protects the acquirer: proper objective information forms the basis for drawing the right conclusions.

With a future extension of the soil certificate, OVAM also aims to broaden people's awareness, particularly on the crucial role of 'soil' or 'land' as a common and the related ecosystem services provided by the soil. OVAM wants to spread the message that the prevention of pollution and the sustainable protection and management of the soil as natural capital provides a healthy living environment and many other benefits to society. Considering land as a common, the owner of a plot needs to consider not only his rights, but also his obligations towards society and the natural environment at large.

Anticipating the future, OVAM also considers the possibilities of an evolution in the provision of information on some soil stewardship principles, e.g. on soil ecosystem services (e.g. the role of biodiversity, …), sustainable soil management, etc. As such, in addition to the general 'the polluter pays' principle, the owner and the beneficiaries jointly take responsibility for their obligations towards society at large.

The extended and continuously evolving soil certificate helps, in a first step, to provide the necessary background information, as well as the motivation for raising people's awareness of these issues. Inform and sensitize, so citizens and entrepreneurs will appreciate a healthy soil and handle it with care.

REFERENCE

McBratney, A. & Field, D.J. & Koch, A. 2014. The dimensions of soil security. *Geoderma* 213: 203–213.

Soil security and research needs

Which R&D needs for a sustainable soil management and land use?

M.C. Dictor & V. Guérin
BRGM, Orléans Cedex, France

S. Bartke
German Environment Agency, Germany

ABSTRACT: Sustainable land management seeks to balance the demand and supply of resources and our natural capital, to cope with the effects of several driving forces putting pressure on the systems and to decrease the global footprint of human made production and consumption. A bottom-up approach engaging about 500 stakeholders from 17 European countries from soil, sediment, water and land-use management communities highlighted more than 1,000 research questions which were clustered according to the 4 themes of the conceptual model of the H2020 INSPIRATION project (Demand, Natural Capital, Land Management and Net Impact). The last step of the analysis allowed aggregating cross-national and cross-sectorial research questions into clustered thematic topics for the 4 research themes mentioned above and of 17 integrated research topics which are identifying theme-overarching research priorities. All the integrated research topics will be the core of the SRA, which will provide a framework for developing, transferring and applying knowledge to tackle societal challenges related to the soil-sediment-water system that will support a resource efficient and low carbon economy.

1 INTRODUCTION

Sustainable land management aims to balance demand and supply between resources and our natural capital in order to cope with the effects of several factors that exert pressure on systems and reduce the overall footprint of production and consumption. How we manage land resources and land management is essential to ensure Europe's transition to a sustainable future for its citizens. Applying research results and knowledge is fundamental to improving our approaches to securing soil and land for future generations, for competitive economies and healthy landscapes. Efficient use of ecosystem services provided by the soil-water-sediment system provides solutions to meet societal needs and societal challenges.

The objective of the INSPIRATION project is to develop a European Strategic Research Agenda (SRA) for soil management and environmentally sound land use that is socially acceptable and economically affordable.

2 CONCEPTUAL MODEL

A conceptual model was develop at the beginning of the project and structured future research needs (Makeschin et al., 2016) (Figure 1). Sustainable

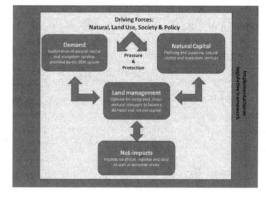

Figure 1. Conceptual model of INSPIRATION research clusters.

land management must seek to balance the demand and the supply, with the latter being base on the resources provided by our natural capital. And the net impact, meaning the local to global footprint of human land management decisions, must be assessed and minimized. Behind the conceptual model arise the question of the conflicting interest regarding land-use among the relevant stakeholders in a society (land planners, citizens, farmers …).

3 METHODOLOGY

The implementation of this SRA must be primarily driven by end-users. INSPIRATION aims to find models for the implementation of the SRA and to prepare a network of public and private financing institutions ready to finance the implementation of the SRA.

The INSPIRATION project is a consortium of 22 partners from 17 countries. It also includes an International Advisory Board of 10 members.

The methodology developed within the INSPIRATION project is a bottom-up approach based on interviews with National Key Stakeholders (NKS) in the public and private sphere, the scientific community, the national funders and the society in each of the 17 countries. The motivation for this process was to ensure that research, development and innovation needs of stakeholders were well identified. The multi-national methodology was built on a multi stakeholders and interdisciplinary approach approached by the National Focal Points (NFP) working as knowledge facilitators. For this consultation, a well-balanced division between 3 main typologies of actors in the thematic area of INSPIRATION, End-Users, Funders and Researchers was done (Figure 2).

According to their professional categories, 1/3 were issued from the research sphere, 1/3 from the public institutions and the last 1/3 from the private and society community (Figure 2).

The bottom-up approach was organized in 3 steps: (i) a desk-exercise to collect relevant informations (reports, publications, ...) which was used to prepare the next step, (ii) individual NKS's interviews using the same questionnaire in the 17 countries (Brils et al, 2015). NKS were chosen for their nationally recognized expertise in their field (Figure 3). NKS were interviewed on research and innovations needs, connecting sciences-policy/practice and about national and international 1

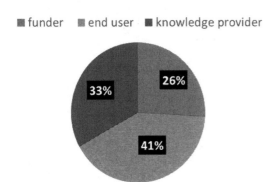

Figure 2. Typologies of national key-stakeholders interviewed by the national focal point.

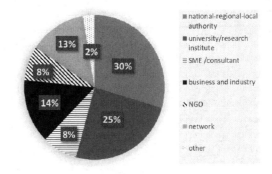

Figure 3. Professional categories of National Key-Stakeholders interviewed by the National Focal Point.

Figure 4. Bottom-up approach for Strategic Research Agenda elaboration.

funding schemes, (iii) NKS's national workshop which was an exchange space for discussion and synthesis of national R&I needs to be raised at a European level (Figure 4).

4 INTEGRATED RESEARCH THEMES

A first step of clustering national research priorities was performed for each of the 4 themes. The national research demands were collated, reviewed and synthetized in order to identify cross country research demands. At the end of this step over 200 research topics and more than 1,000 research questions were identified. At the end of the clustering of national research questions by theme, some of them addressed several themes. So, 17 integrated research themes (IRT) were created in order to put research issues in perspective to societal challenges and stimulate creation of transnational funding programs (Nathanail et al, 2017):

IRT-1: Integrated Environmental Assessment and Soil Monitoring for Europe

IRT-2: Recognizing the values of ecosystem services in land use decisions

IRT-3: From indicators to implementation: Integrated tools for a holistic assessment of agricultural and forest land use
IRT-4: Bio-Economy—unleashing the potentials while sustaining soils
IRT-5: Integrated scenarios for the Land-Soil-Water-Food system under societal pressures and challenges
IRT-6: Indicators for assessing the efficiency of the Soil-Sediment-Water-Energy system of resources
IRT-7: Farming systems to maintain soil fertility while meeting demand for agricultural products
IRT-8: Circular land management
IRT-9: Policies to effectively reduce land consumption for settlement development
IRT-10: Stakeholder participation to facilitate the development of liveable cities
IRT-11: Integrated management of soils in urban areas
IRT-12: Environmentally friendly and socially sensitive urban development
IRT-13: Urban Metabolism—Enhance efficient use of soil-sediment-water resources through a closing of urban material loops
IRT-14: Emerging contaminants' in soil and groundwater—ensuring long-term provision of drinking water as well as soil and freshwater ecosystem services
IRT-15: Sustainable management to restore the ecological and socio-economic values of degraded land
IRT-16: Innovative technologies and eco-engineering 4.0: Challenges for a sustainable use of agricultural, forest and urban landscapes and the SSW system
IRT-17: Climate change challenges—improving preparedness and response for climate conditions and related hazards

5 CONCLUSIONS

Increasing societal challenges facing Europe require research and innovation which integrates different approaches from across research disciplines. Toward the elaboration of the European Strategic Research Agenda on land and Soil-Sediment-Water system management, transnational research themes were identified on the basis of transnational stakeholder's consultation. From these results, a cross-border and interdisciplinary dialogue will be organized between the user groups concerned, private and public funding institutions and scientific communities in Europe in order to achieve a transnational, hierarchical SRA with a model dissemination of the latter.

ACKNOWLEDGMENT

The INSPIRATION project for 'INtegrated SpatIal PlannIng, Land Use and Land Management Research AcTION' is a three-year Coordination and Support Action of the H2020 program (2015–2018) coordinated by the German Environment Agency.

INSPIRATION acknowledges the received funding from the European Community's HORIZON2020 Framework Programme under grant agreement no. 642372.

REFERENCES

Brils J., Maring L., Darmendrail D., Dictor M.C., Guerin V., Coussy S., Finka M., Bal N., Menger P., Rehunnen A., Zeyer J., Schroter-Schlaack C., Villeneuve J., Gorgon J., Bartke S. 2015. Delivrable D2.3 of the projet INSPIRATION. UBA: Dessau-Roblau, Germany.

Makeschin F., Schroter-Schlaack C, Glante F., Zeyer J., Gorgon J., Ferber U., Villeneuve J., Grimski D., Bartke S. 2016. INSPIRATION report concluding 2nd project phase: Enriched, updated and prioritized overview of the transnational shared state-of-the-art as input to develop a Strategic Research Agenda and for a matchmaking process. Public version of the final version as of 30.10.2016 of deliverable D3.4 of the HORIZON 2020 project INSPIRATION. EC Grant agreement no: 642372, UBA: Dessau-Roblau, Germany.

Nathanail C.P., Boekhold A.E., Bartke S., Grimski D. 2017. The Europan's Strategic Research Agenda for Integrated Spatial Planning, Land use and Soil Management-June 2017 Green Paper, HORZON 2020 project INSPIRATION. EC grand agreement no 642372, UBA: Dessau-Roblau, Germany.

Conclusion

The 2nd global soil security conference—conclusions and prospects

Anne C. Richer-de-Forges & Dominique Arrouays
INRA, InfoSol Unit, Orléans, France

Florence Carré
INERIS, Verneuil en Halatte, France

Johan Bouma
Wageningen University & Research, Wageningen, The Netherlands

Alex B. McBratney
The University of Sydney, Sydney NSW, Australia

ABSTRACT: The 2nd Global Soil Security conference, assembled in Paris on December 5th and 6th 2016, aimed at connecting businesses, practitioners, policymakers and researchers to discuss soil security focusing particularly on best working practices, business solutions, and international initiatives. Although good practices aimed at achieving soil security were presented during the conference, it appeared there are still crucial operational and scientific targets to meet in order to reach a fully operational system to achieve soil security. These targets include:

i. improved communication with actors dealing with the practicalities of soil management and conservation issues;
ii. more interdisciplinary activity between soil science and the Humanities and Social Sciences;
iii. more attention for the communicative power of modern social media, focusing on the soil-water-plant-atmosphere system and its major impact on human life, and
iv. more attention for the role of soils in defining ecosystem services and Sustainable Development Goals
In the following decade, the five C's of Soil Security can provide a pragmatic guideline to achieve sustainable land use. In this context, soil connectivity should receive a major thrust when planning research activities.

1 INTRODUCTION

The five Soil security dimensions (capability, condition, capital, connectivity and codification) (McBratney et al. 2014; 2017) were first presented and discussed during a regional Agricultural Research Symposium in Sydney in 2012. The first Global Soil Security conference was held in College Station (USA) in May 2015. The second global conference was held in Paris on December 5–6, 2016 and was focused on connecting businesses, practitioners, policymakers and researchers by presenting best working practices, business solutions, scientific outcomes and international initiatives, all of them contributing to the protection and sustainable management of soils. After introductory talks from the French Minister for Agriculture and from the Directors General of INRA and INERIS, the conference presented 44 oral presentations, 16 poster presentations and round-table panel discussions that this chapter aims to summarize.

2 COVERAGE OF THE FIVE DIMENSIONS OF SOIL SECURITY

The way in which the five soil security dimensions (capability, condition, capital, connectivity and codification) were covered during the conference was analyzed by ranking all papers on a 0 to 5 scale according to their relevance to each of the five dimensions. Then, a mean value of all papers was calculated, as shown in Figure 1.

Soil capability and condition were most frequently addressed, followed by capital, codification and, finally, connectivity. Indeed, capability and condition assessments are by now well established within soil science based on widely available knowledge and expertise produced from field experiments and theoretical approaches. This includes increasingly important digital mapping and automated monitoring of soil processes. Studies of soil capital receive only half as much attention, but a recently observed increase is due to observed

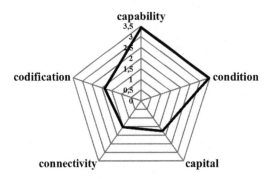

Figure 1. Coverage of the five dimensions of soil security during the 2nd Global Soil Security conference, Paris, France, 5–6 December 2016.

economic benefits of permaculture or of modern agro-ecological approaches in farming systems. The decrease of soil biodiversity and potential action feedbacks on climate change and soil health call for more knowledge on valuing soil capital, for example by soil carbon stock assessment, or, more broadly, by valuing ecosystem services rendered by soils. Connectivity and codification received particular emphasis due to the diversity of the audience and the welcome presence of policy makers at this conference. This has certainly contributed to increase in the perception and knowledge of connectivity, at least in the French and European contexts. However, as stated earlier, codification and connectivity received about half the attention compared of capability and condition, implying that soil science needs to be better linked with the Humanities, Social and Political Sciences.

3 ARE SOME KEY CONNECTIONS MISSING?

As emphasized in the last section, more attention is needed for connectivity. The different disciplines that should be actively involved are sociology, psychology and/or anthropology. Increasing the knowledge on codification, which is also currently receiving less emphasis, requires attention for the role of soils in national and international environmental law and private-sector certification. Particular attention for risk management science would appear to be highly relevant allowing short term versus long term risk assessment procedures in developing land use programs. Developing socially acceptable decision criteria need special attention. Finally, the need to quantify the dimensions of soil security by assigning values or indices was stressed many times during the conference. A quantitative assessment framework is a key priority.

This conference provided an opportunity to gather scientists working on both agricultural and on industrially polluted soils. They shared experiences on the transfer and behaviour of chemical substances within and between soils, including possible human impact. More contact between these scientists are being planned following this inspiring meeting. Discussions also emphasized lack of attention for the soil-air interface. Aside from widely studied greenhouse gas emissions, more attention is needed of wet and dry deposition processes. Also under some climatic and soil conditions, wind erosion can lead to huge amounts of dust that can be blown over thousands of kilometres. Thus, soils can strongly affect air quality and vice-versa.

4 ARE THERE OUTSTANDING SCIENTIFIC QUESTIONS?

Several questions were raised during the conference that require additional attention. For example:

- to which extent does soil biodiversity make a difference for the 5 dimensions? If it does, can we really quantify this?
- when judging soil performance, what is the baseline level? Does it make sense to consider the concept of a pristine, virgin soil when none exist anymore?
- is biochar only a fashionable topic? Does the general use of biochar offer a realistic perspective?
- is the concept of soil security perhaps too anthropocentric?
- how to quantify ecosystem services, environmental and economic benefits for different goals and at different scales? Are the soil quality and soil health concepts helpful in this context?
- how to tackle ethical issues when considering soil management?

5 CONSENSUS

All participants agreed on the following key points:

- Soil security is crucial and at the centre of global issues (food, water, energy, climate, biodiversity).
- Practical examples of soil management solutions exist like increasing crop rotation length and diversity, conservation tillage, better water management, contaminated sites remediation. However, one size does not fit all and solutions need to be adapted to the diversity of soils, climate, land use and agricultural systems. That is why, we need more active communication on potential solutions in different contexts that

should be taken up in more efficient policies. This communication should be adapted to the audience. If so, it can improve soil connectivity;
- We need progress on how to measure and map soil security;
- We need progress on how to manage soil security;
- We need to propose solutions to bring soils from condition to capability. (This is the main raison d'être of all soil scientists.)
- We need global instruments (treaties, conventions, accreditation) on soil security. There is some progress, but soil is still sometimes hidden behind other issues.
- We need local actions to implement soil security (from national to local policies, to farmers and to citizens).
- We need instruments and tools at different scales: local (e.g. farm management certification, tools for monitoring, tools for remediation, standards, key soil condition indicators within different landscapes), regional/national (e.g. land-use planning, regulations, market, branding), global (e.g. treaties, conventions).

6 CONCLUSIONS

The conference has shown that bringing together members of the business community, practitioners, policymakers and researchers, with the objective of exploring the potential of the soil security concept, can really contribute to increase knowledge and understanding of protection, sustainable management, and potentially, enhancement of soils. In this context, the soil security concept with the 5C's is quite valuable. However, there are still crucial operational and scientific targets to be met before a fully operational soil security paradigm can be launched. As stated, more and better communication is needed between the wide group of actors concerned with soil management and conservation. More effective interdisciplinarity between soil science and the Humanities, Social and Political Sciences is very important as is the interaction with modern social media and communication sciences. In terms of the soil security concept, this implies more emphasis particularly on the connectivity aspect. Potentially this is where the greatest gain may be made for a modest effort.

REFERENCES

McBratney AB, Field DJ, Koch A. 2014. The dimensions of soil security. Geoderma; 213:203–213.
McBratney, AB, Field, DJ, Morgan, CLS, Jarrett, LE. 2017. 'Soil Security: A Rationale' Global Soil Security, pp. 3–14.

Author index

Angevin, F. 65
Antón, R. 79
Antoni, V. 33
Arias, N. 79
Arrouays, D. 43, 133

Back, P.E. 99
Bagnall, D.K. 19
Bailey, S. 57
Bal, N. 119
Bartke, S. 127
Bescansa, P. 79
Bispo, A. 33
Bouma, J. 3, 133
Bryce, A. 11

Campillo, R. 79
Carré, F. 133
Cavan, N. 65
Ceenaeme, J. 119
Cousin, I. 65

Dictor, M.C. 127

Eglin, T. 33
Enell, A. 99
Enrique, A. 79

Feix, I. 33
Field, D. 11, 91
Finnell, P. 57
Fort, J.-L. 33
Franco, S. 107

Gimona, A. 47
González, J. 79

Grundy, M. 43
Guérin, V. 127

Hempel, J. 43
Hernández, I. 79
Honeycutt, C.W. 113
Hooimeijer, F.L. 73

Jones, N. 57
Jones, S.R. 113

Kome, C. 57

Labreuche, J. 65
Lene, J. 57
Libohova, Z. 43, 53, 57
Lindbo, D. 53

Marinari, S. 107
Maring, L. 73
Matos, M. 57
McBratney, A.B. 11, 43, 133
McIntosh, Wm.A. 19
McKenzie, N. 43
McVey, S. 57
Minasny, B. 43
Montanarella, L. 29
Monteith, S. 57
Morgan, C.L.S. 19
Moscatelli, M.C. 107
Moyce, M. 11

Norrman, J. 73

Ohlsson, Y. 99
Orcaray, L. 79

Panagos, P. 29
Poggio, L. 47

Rath, B.F. 113
Rayé, G. 33
Reich, P. 57
Richer-de-Forges, A.C. 43, 133
Rolfes, T. 57
Roudier, P. 43

Sauter, J. 33
Scheffe, L. 57
Schoeneberger, P. 53
Seybold, C. 53
Shafer, S.R. 113
Slak, M.-F. 33
Soubelet, H. 33
Southard, S. 57

Thorette, J. 33

Van Den Driessche, W. 119
Van Gestel, G. 119
Virto, I. 79

Wills, S. 53
Wissocq, A. 65
Woodward, R.T. 19
Wysocki, D. 53